Charles Pritchard

Uranometria Nova Oxoniensis

A photometric determination of the magnitudes of all stars visible to the naked eye

from the pole to ten degrees south of the equator

Charles Pritchard

Uranometria Nova Oxoniensis
A photometric determination of the magnitudes of all stars visible to the naked eye from the pole to ten degrees south of the equator

ISBN/EAN: 9783337406394

Printed in Europe, USA, Canada, Australia, Japan

Cover: Foto ©Andreas Hilbeck / pixelio.de

More available books at **www.hansebooks.com**

URANOMETRIA NOVA OXONIENSIS

A PHOTOMETRIC DETERMINATION

OF THE

MAGNITUDES OF ALL STARS VISIBLE TO THE NAKED EYE
FROM THE POLE TO TEN DEGREES SOUTH OF
THE EQUATOR

BY

C. PRITCHARD, D.D., F.R.S., F.G.S., F.R.A.S.

SAVILIAN PROFESSOR OF ASTRONOMY IN OXFORD

Orford

AT THE CLARENDON PRESS

M DCCC LXXXV

London
HENRY FROWDE

OXFORD UNIVERSITY PRESS WAREHOUSE

AMEN CORNER, E.C.

INTRODUCTION.

General Historical Survey of Astrometry.

In the present volume is condensed the record of three years' labour on the Photometry of the stars visible to ordinary and unaided eyes, from the North Pole, to about ten degrees South of the Equator. Estimations of the relative brightness of these stars have been made from time to time by various astronomers, but by none more successfully than by the illustrious Argelander, who gave the results of his survey in the *Uranometria Nova*, published at Berlin in 1843. It may not be too much to say of these estimations that, taken as a whole, and viewed as estimations made without the aid of graduated instruments, they are not likely to be surpassed in point of precision; yet it is not any disparagement of their intrinsic value to add, that the requirements and progress of modern astronomy demand greater exactitude than that which Argelander's work justly claims, and a precision exceeding any that can be expected from observations made with the unaided eye.

In the place of these estimations, I now propose to substitute instrumental measures, made at the University Observatory in Oxford, with the aid of a Photometer devised for the purpose by myself, and in which a long and varied experience warrants me in placing a high degree of confidence.

From time immemorial, reaching probably even far beyond the Homeric epoch, the configuration of the stars in the heavens served, mainly through their risings and settings, as a rough calendar for the regulation of the dates of civil and religious proceedings, and for the purposes of agriculture and navigation. There is a record that such configurations were depicted on a Celestial Globe, constructed by Eudoxus four centuries before the Christian Era; and there can be no doubt that some method was at the same time devised for the designation thereon of their relative brightness. A copy or modification of this ancient Globe, supported on the shoulders of a marble Atlas, was dug up from the ruins of Rome, and now furnishes one of the most interesting objects of antiquity in the Royal Museum at Naples, placed there by the munificence of Cardinal Farnese. It may be mentioned in passing that the configurations of the constellations on this Globe are substantially the same as those recognised at the present day.

But it is to Ptolemy, in his immortal work the Μεγάλη Σύνταξις, or Almagest as it was termed by his Arabian translators, that we are indebted for a record, not only of the celestial co-ordinates of the stars visible in his day (cir. A.D. 150), at Alexandria, but also for a catalogue of their relative brightness, such as he had himself probably received from Hipparchus and his predecessors more remote.   It is a remarkable instance, among many others, of the incisive intellect of the ancient Greeks, that they adopted not only an admirable nomenclature for stellar brightness, which has remained substantially unaltered to the present day, but one which even in its minuter sub-divisions has been but slightly improved by modern astro- nomers.   It is still more remarkable that in this ancient and conventional nomenclature, they practically but unconsciously anticipated an important and fundamental law in Photometry, the first verbal expression of which was brought into prominence by Fechner at so recent a date as 1859[1].

These ancient astronomers, as is well known, divided the brightness of the stars, conceived by them under the thought of 'Magnitude' (Μέγεθος) into six classes.   They assigned the 'first magnitude' to a small group of the brightest stars, and then proceeded step by step in successive groups to the sixth, which included all stars shining with the feeblest lustre admitting of appreciation by the naked eye.   There was again in Ptolemy's catalogue a sub-division of each magnitude into three, an amount of precision which seems to have been subsequently abandoned for a long period by his successors.   It was, however, a nomenclature resumed by Flamsteed and adopted by Argelander in his *Uranometria Nova*, and perhaps it is not too much to say that a finer or more delicate sub-division of stellar lights, than that denoted by the third of a magnitude, is not readily, and by direct means, ordinarily appreciable by the human eye.

These tabulated magnitudes of individual stars, recorded in the catalogues of Ptolemy, remained practically unimproved from his day to that of the elder Herschel at the close of the eighteenth century. Nevertheless the majority of those who flourished in the long line of eminent astronomers between these distant intervals, did make some few and feeble attempts to improve estimations, which they could not do other- wise than feel, were not more than provisional.   Among these the most honourable place must be assigned to Abd-al-Rahman Al Sûfi, who, about the year 930, re-examined Ptolemy's work by a comparison with the heavens[2]. Tycho (cir. 1570) made no advance herein; the same remark applies also to Hevel (cir. 1680).   It might have been expected that Bayer, when, in A.D. 1601, he bestowed on astronomers the memorable boon of a new nomen- clature of the stars in their several constellations, through the application to them of the letters of the Greek alphabet, would have availed himself of the golden opportunity thus afforded him for a re-examination of the ancient

[1] Über ein psychophysisches Grundgesetz und dessen Beziehung zur Schätzung der Sterngrössen, von G. Th. Fechner.   Leipzig, 1859.
[2] Description des Étoiles fixes, composée au milieu du dixième siècle de notre ère par l'astronome persan Abd-al-Rahman Al Sûfi, St. Pétersbourg, 1874 ; and Monthly Notices, vol. xlv, p. 417.

magnitudes.　Unaccountably, Bayer allowed the opportunity to pass, and thereby laid himself open to the caustic remark of Delambre, as to the cheap rate at which he had acquired immortality.

Finally, Flamsteed (cir. 1689), the first astronomer who applied the telescope to systematic celestial measurements, re-introduced the sub-divisions of a magnitude into thirds, by means of the notation now in general use : but in other respects exhibited either negligence or unconcern in his estimations of relative stellar brightness.　It was this disregard of precision on Flamsteed's part which mainly induced Sir William Herschel [1] to turn his own attention to the subject.

Penetrated with the importance of some record whereby obvious variations in the brightness of stars could be ascertained ; important, not solely on the side of inevitable curiosity, but from the consideration that our own star, the Sun, might itself be variable in light and heat, William Herschel (cir. 1796) set himself to the task of forming that record, after his own manner and from his own resources.　With this view, he did not propose to verify or improve Ptolemy's magnitudes, and still less those of Flamsteed, but he adopted a more practical expedient, leading, as he rightly thought, both to greater exactness and to greater utility in the direction of the variability intimated above.　Accordingly, he divided a constellation into small groups of two, or three, or four stars of nearly equal brightness, and he then arranged the stars in these small groups in the order of their lustre.　One star might in this way be found in more groups than one, and thus might furnish the means also of connecting several groups together.　Beyond this, he attempted with much success to designate the degrees of the various differences in lustre, not by numerical sub-divisions of magnitude as had heretofore been the expedient, but by the introduction of symbols such as dots, commas, semicolons, &c., placed between the stars whose brightness was compared.　For instance, in his nomenclature, a dot placed between two stars would indicate all but absolute equality in brightness, the second star however being, if anything, somewhat less bright than the star which preceded the dot.　A comma between them would indicate a somewhat greater difference of brightness, and so on.　It is due to the reputation of this great practical astronomer to state that, on a photometric examination of the small differences of light indicated by some of these symbols, there is a precision and a consistency generally observable, which excite admiration.

It was evidently no part of Sir William Herschel's intention to form, or even to lay the foundation for forming a systematic catalogue of the brightness of the stars; but, as already intimated, it was his intention to record the means of detecting any, the slightest, variation which might occur at future periods in any of the stars forming any one of his groups. These groups or short sequences are so numerous that they have not unnaturally induced some astronomers to attempt the deduction therefrom

---

[1] Philosophical Transactions, 1796, p. 166.

of a complete catalogue of brightness systematically and numerically arranged. Such an unavailing attempt has been made by the author of the present volume, and, could it be successful, the value as a record would be great indeed. But the groups are often so unconnected, the lacunæ are so numerous, and although the estimates of light between stars in the same group are for the most part unimpeachable, nevertheless these considerations render a complete catalogue of magnitudes, considered as Herschelian, unattainable. The danger consists in the liability of importing so much from the observations, whether photometric or otherwise, of other astronomers, as would (and, in fact, do) render the final result, not so much Herschel's unmixed production, as the reflection therein of other catalogues formed by these astronomers. It must be here repeated, that these remarks are not meant as the slightest disparagement of the illustrious astronomer's work, for in this matter nothing was probably less in his mind, than the formation of a systematic record of star magnitudes.

Sir John Herschel, however, during his busy life at the Cape of Good Hope (1835–1838) did, in his mind, propose to complete the Astrometry of the Heavens so far as was visible to the naked eye. His object appears to have been, through loyalty to his father's memory, and for the permanent advance of Sidereal Astronomy, to apply to the Southern Heavens that same sort of scrutiny on which the latter had so long laboured at Slough, with regard to the Northern. The process which he adopted was to divide the stars visible to the unaided eye, into many sets of long and interlacing sequences arranged in graduated lustre. These, when completed, he considered would furnish him with a systematic catalogue of magnitudes, possessing far greater accuracy than any that had heretofore existed. This plan, it will be observed, was generically different from that adopted by the elder Herschel, whose aim was confined to the comparison of stars divided into small sets of nearly equal brightness. While Sir John Herschel was engaged in this project, and in the midst of others of greater magnitude, he invented a photometer, by means of which he hoped to arrive instrumentally at the same sort of results which heretofore he and his predecessors had sought from estimations alone, though now with far greater precision. The implement he devised was of rude construction, such as could be arranged from the scanty resources of a distant colony, but it was, in principle, sufficient for his purpose. Roughly described, it consisted of a pole, a prism, a small lens of short focus, a few strings and a graduated tape. With these materials properly arranged, he could obtain, in the focus of the lens, a microscopic image of the moon, and this he could view in any direction, and at any measured distance from the eye, so that being brought into the same line of sight with any particular star, he could alter the distance of the tiny image, until it and the star appeared to be equally bright. In this way, the brightnesses of some sixty-nine stars were compared with that of α Centauri, and the results were tabulated. These relative light intensities, and the results of the sequence-observations were then expressed in magnitude, in such a way and on such a scale as best to accord with the accepted magnitudes contained in the best catalogues then extant.

The details of the process, at once ingenious and complicated, can be mastered only by a reference to the original account given by the Author in *The Cape Observations.* Unfortunately the entire project was not completed. An interesting comparison of some of the results with those obtained by the Oxford Photometry will be found in Vol. XLVII of the *Mem. R. A. S.*, and more elaborate descriptions will be found in the *Photometric Researches at Harvard College* both by Mr. C. S. Peirce and by Professor Pickering.

About the same time that Sir John Herschel was engaged in his astronomical observations at the Cape of Good Hope, Argelander was at work at Bonn, on the similar work of Astrometry; but fortunately, with this difference, that he was able to complete it. In 1843 he published his *Uranometria Nova,* containing the estimated magnitudes of all the stars visible to the naked eye in central Europe. This, in fact, is the first successful attempt made by modern astronomers to arrange in an original and independent catalogue, the relative brightness of the stars, and it must ever remain a striking instance of what can be achieved by well-directed perseverance and accurate discernment. Argelander has not recorded the details of the method which he adopted to secure his results, nor is it possible to say by what means or mental impressions, he preserved a fair uniformity of scale, and a general conformity to the magnitudes recorded by his predecessors. It may be sufficient to say that some not inconsiderable variations of light-ratio do occur, depending on magnitude; but such variations are unavoidable, when the scale is the result of mental impressions, rather than an instrumental measure.

Argelander's work was soon followed by another of still greater magnitude and importance in which he, with the most able assistance of Drs. Schönfeld and Kreuger, has recorded the approximate celestial co-ordinates and the magnitudes of no fewer than 324,000 stars. In this case the magnitudes are such as they are estimated to be, when seen through the telescope, and not, as in the *Uranometria Nova,* observed with the eye alone. If there are found some slight variations both in the magnitude of individual stars, and in the light-ratios existing in the general estimate of the larger intervals, the cause is to be sought in the same sources of imperfection as those just referred to in the case of the *Uranometria Nova.*

More recently, the heavens have again been scrutinized by Heis, at Münster, who has re-examined Argelander's *Uranometria Nova* and has added a considerable number of stars of a fainter lustre; but many of these, it is not too much to say, are beyond the vision of ordinary eyes. Houzea, also by observations made during a residence for that purpose at Jamaica, has rendered good service to astronomy, by publishing the results of his own estimations. The same remark applies also to the astrometry of M. Flammarion at Paris. Dr. Gould has recently enriched the resources of astronomy by his astrometry of the southern heavens, with a degree of excellence such as to have been crowned by the bestowal of the Gold Medal of the Royal Astronomical Society.

Instrumental Aids to Sidereal Photometry.

Contemporaneously with Herschel's invention of the rude, but ingenious photometer at the Cape, Steinheil at Munich was arranging a more refined, if not more accurate instrument, which was subsequently employed by Seidel. It consisted of a small telescope, with a divided object-glass, each of the halves of which was furnished with a reflecting prism, so that by means of suitable mechanism, the images of two stars might be viewed side by side in the telescope. The two halves of the object-glass were capable of motion in the direction of the axis of the telescope, so that the images of the two stars, when diffused into small discs, might be equalized in brightness, by placing the respective object-glasses at different measurable distances from the eye-piece. In this way, the intensities of a considerable number of the brighter stars were compared. Seidel, however, did not reduce his results to magnitude, but left them as logarithms of light, compared with that of Vega, considered as unity. It is difficult to correctly estimate the precision of Seidel's results, but an interesting comparison of his measures, set side by side with Herschel's, will be found in Vol. XLVII of the *Roy. Ast. Soc. Memoirs.* Seidel also used his instrument successfully, in measuring the amount of light lost by a star, through absorption in passing through the Earth's atmosphere. Practically he verified Bouguer's results obtained in 1725; he expresses some surprise at the comparative exactness of the verification: but subsequent researches, and among others those of the author of this volume, have confirmed the substantial accuracy of their determinations.

Zöllner also devised a Photometer in which he successfully applied Arago's suggestion[1] of the method of polarization. The comparison star in Zöllner's instrument was formed artificially by means of a lamp, the light of which could be reduced by double refraction through a measurable quantity, until it was judged to have the same intensity as that of any required star, in the same field of view as the artificial star itself. This form of photometer has been in considerable use, chiefly by Zöllner[2] himself, by Pierce[3], Wolff[4], Müller[5], Lindemann[6], and others. The unavoidable want of uniformity in the light of the lamp, and the impossibility of imitating exactly the appearance of an actual star, are fundamental difficulties in the use of this otherwise convenient and ingenious instrument. Zöllner considered that he could imitate the colour also of a star, by the intro-

---

[1] Œuvres de Fr. Arago, Tome x, p. 184.
[2] Grundzüge einer allgemeinen Photometrie des Himmels, 1861.
[3] Annals of the Harvard College Observatory, vol. ix.
[4] Photometrische Beobachtungen an Fixsternen. Bonn, 1875.
[5] Publicationen des Astrophysikalischen Observatoriums zu Potsdam, Nr. 12.
[6] Helligkeitsmessungen der Bessel'schen Plejadensterne.

duction of suitable polarizing apparatus, between his lamp and the eye-piece of the telescope ; but the success of such an arrangement is doubtful.

An improvement both on Seidel's and Zöllner's instruments has been devised by Professor Pickering at Harvard, to which he has given the name of Meridian Photometer, and in which he dispenses with the divided object-glass of the former, and the lamp of the latter. There is a telescope tube placed horizontally at right angles to the meridian in which is placed two object-glasses, with their axes slightly inclined to each other. Each object-glass is 4 centimetres in diameter, and is armed with an adjustable reflecting prism. By these means, images of Polaris and of any star near the meridian, may be formed in the common focal plane of the two object-glasses, after having passed through a suitable combination of double refracting prisms ; and then polarized images of the two stars may be equalized, and their brightness ascertained. For a fuller description of the instrument recourse must be had to the original memoir of its inventor forming Vol. XIV of the *Annals of the Astronomical Observatory of Harvard College*. The construction is somewhat intricate, but if it be practicable to bring all the light falling on the two object-glasses, effectively through the double refracting arrangements, into the pupil of the eye, and then to afford adequate time for the necessary repetitions of the equalizations, then this form of photometer seems to offer very great advantages in stellar photometry. By means of it, Professor Pickering and his assistants have produced a catalogue of photometric measures, of all stars reputed to be equal to or brighter than the sixth magnitude, visible at Harvard, for which the permanent gratitude of astronomers will be accorded. Professor Pickering has also applied the principle of polarization of light to the photometry of double stars with success ; but the question of the equalization of some of the more vividly coloured stars still remains of doubtful solution, and it may be is beyond the reach of physical considerations alone.

The method of photometry by means of varying the telescopic aperture has received a new and practical development at the hands of Mr. Knobel. By using the mirrors of a Newtonian telescope, one or both of them in an unsilvered condition, he has contrived, on what appear to be unimpeachable conditions, to extinguish the light of stars, well visible to the naked eye, and thereby is enabled to compare their relative magnitudes [1].

Description of the Wedge Photometer.

The instrument with which the researches in this volume were made, differs generically from all the preceding photometers, unless indeed the eye itself be regarded as therein displacing or virtually performing the functions of Zöllner's lamp and artificial star of comparison. It is constructed on the principle that light in passing through a transparent homogeneous medium, loses an amount of intensity depending on the thickness

[1] Monthly Notices of the Royal Astr. Soc., vol. XXXV, p. 100.

of the latter. Hence it is easily shown, that if L represent the intensity of a small pencil of light incident perpendicularly on a medium bounded by parallel plane surfaces, such as a rectangular prism of neutral tinted glass of the thickness τ, it will emerge as a similar pencil with an intensity L', such that

$$\text{Log}\,\frac{L}{L'}= K\tau\,; \qquad\qquad (1)$$

where K is some constant dependent on the material of the glass and the nature of the light.

If then in passing from any one 'magnitude' to the next fainter, we adopt a constant light-ratio of (ρ), and if L_n, L_{n+x} represent the intensity of two lights whose magnitudes are n and $n+x$, forming the incident and emergent pencils, we shall have

$$\text{Log}\,\frac{L_n}{L_{n+x}} = K\tau \text{ or } \text{Log}\,\rho^x= K\tau \text{ or } x = \frac{K}{\text{Log}\,\rho}\,\tau, \qquad (2)$$

i.e., *the light lost in passing through the medium, measured by the alteration of 'magnitude,' is proportional to the thickness of the absorbing medium through which it has passed.*

The above is the fundamental principle on which the wedge photometer is constructed. This general principle had been applied by Mr. Dawes[1] and by others, but owing to various circumstances of misconception, and the unavoidable difficulties in the earlier manipulation, no definite or systematic results were then obtained.

The wedge photometer as employed at Oxford is a wedge of very nearly neutral tinted glass GDF, six and a half inches long, an inch broad, and 0.145 inch thick, tapering off to D, where it is 0.02 inch. Cemented to it is a similar wedge GDH, of white glass, placed the reverse way. The whole forming a rectangular prism. This glass prism or '*wedge*' as it will henceforth be called, is enclosed in a brass cellular rim with bevilled edges, one of which is divided into tenths of inches, the divisions being distinct and white for visibility at night. It slides in a groove on the brass cap of the eye-piece of any telescope, close to the achromatic eye lens, and is thus placed between the eye of the observer and the telescope. In the focus of the eye lens is a diaphragm, pierced with a number of small holes, which vary from the hundredth of an inch to a quarter of an inch in diameter, and in which small circular hole the telescopic image of a star is carefully placed, and there viewed through the wedge. Further, the eye of the observer is directed along the axis of the lens and of the telescope, by means of an external eye hole, placed close to the wedge, and varying from one-twelfth of an inch to a quarter of an inch. This direction of vision is important. A fiducial mark is drawn on the brass cap of the eye-piece, so that the position of the wedge can be distinctly marked and recorded *when the image of the star is just extinguished by the wedge.* Usually the position of the wedge, when the light of the star is just extinguished, is

[1] See Memoirs of Royal Astr. Soc., vol. xxi, p. 557.

observed five times, and the mean of all the five readings is called the '*wedge reading.*' The extinction of a second star is then observed in a similar manner, and the difference between the two wedge readings is called the '*wedge interval.*' This wedge interval is obviously a measure of the difference of the thicknesses of the neutral tinted glass of the wedge at the points where the two stars are respectively extinguished; and it will soon be shown that this 'wedge interval' is also a direct measure of the difference of the '*magnitudes*' of the two stars whose lights are respectively just extinguished at the two points on the wedge.

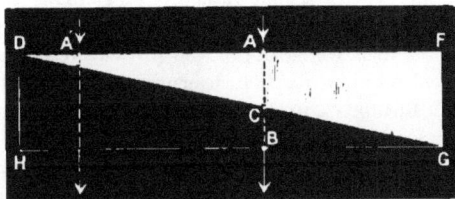

Suppose M_1 to be the magnitude (expressed as a number) of the star's light incident at A, M_1' and M_1'' the magnitudes of these lights as reduced by absorption in passing to C and B respectively. Then from the preceding considerations

$$M_1' - M_1 = K \cdot AC,$$

$$M_1'' - M_1' = K' \cdot CB,$$

hence $M_1'' - M_1 = K'(AB - AC) + K \cdot AC = K'AB + (K - K')AC$

$\qquad\qquad = \text{constant} + K''AD$ where K, K', K'' are constants.

Next suppose the light of a second star incident at A' to be viewed through the wedge and of which the incident magnitude is M_2, then as before

$$M_2'' - M_2 = \text{constant} + K''A'D.$$

In the actual case of moving the wedge to such positions that the stars are just extinguished M_1'' and M_2'' is the limit of vision of faint lights by the observers, so that $M_1'' = M_2''$ and consequently

$$M_2 - M_1 = K''(AD - A'D) = K''AA'.$$

This is the fundamental equation of the wedge, expressing the law and the amount of its action on light. It also follows, as already stated, that if the wedge interval AA' which corresponds to a reduction of light by one magnitude, is found, then the constant K'' is the reciprocal of the number of inches in AA'. Or, in fact, if the wedge interval corresponding to the reduction of any given number of magnitudes or portions of a magnitude could be found, this would also give the value of the constant.

The first method which occurred for finding this constant was the presumed property of light, that if the linear aperture of an object-glass be halved the light of a star would be reduced accurately and exactly to one-fourth of its light, when viewed with the full aperture. In order to exhibit this constant in a convenient numerical form, it is necessary to fix upon some numerical value for the light-ratio (ρ). Several values have been proposed, among others (2) was proposed by W. Struve[1] as, in his judgment, fairly representing the sort of light-ratio which may be detected in the ordinary estimations of magnitude. A much better ratio was proposed by Mr. Pogson[2], who in 1852 assisted Mr. Johnson at the Radcliffe Observatory, viz. $\rho = 2.512 \ldots$ a number whose logarithm is .4, and this ratio being at once convenient and fairly representing the ordinary tabulated magnitudes, meets with a general acceptance among astronomers.

Now the result of an immense number of measures made with one of the two wedges (designated as wedge A) in the present research was, that the mean wedge interval required for the absorption of the light of one magnitude is 0.514 inch. But for various reasons, chiefly because an object-glass is not uniform, either in thickness or in material, I was not satisfied with the strict accuracy of the result. Accordingly, I proceeded to examine the wedge A, (and subsequently the wedge B) by means of double refraction, or polarization of light, as explained in Vol. XLVII *Mem. Roy. Astr. Soc.* The wedge was examined for the amount of absorption at intervals of every tenth of an inch from the one end to the other. A very slight, but not altogether insensible, want of uniformity was discovered in the wedge itself, and the inexactness of the method of apertures referred to above was confirmed. In the *mean* it was found that the wedge interval corresponding to a difference of absorption amounting to one magnitude on Pogson's scale, was 0.539 inch, instead of 0.514 inch. As the result of a very extensive series of comparisons the following table was drawn up. From this table it is to be inferred, that at one inch from the thinner end, the difference of absorption at the beginning and end of the inch amounts to 1.93 magnitude, instead of the mean amount 1.896. At two inches from the end, the difference of absorption for an interval of two inches is 3.81 mag., instead of twice the mean amount or 3.79 mag., and so on.

[1] Mensurae Micrometricae, p. xlii.
[2] Radcliffe Observations, vol. xv, p. 297.

TABLE I.

For Wedge A (6½ inches long).

Wedge Scale Reading.	Magnitude Absorbed.	Wedge Scale Reading.	Magnitude Absorbed.	Wedge Scale Reading.	Magnitude Absorbed.	Wedge Scale Reading.	Magnitude Absorbed.
inches.		inches.		inches.		Inches.	
0.0	0.00	1.5	2.88	3.0	5.69	4.5	8.43
0.1	0.20	1.6	3.07	3.1	5.88	4.6	8.60
0.2	0.40	1.7	3.26	3.2	6.06	4.7	8.77
0.3	0.59	1.8	3.44	3.3	6.25	4.8	8.95
0.4	0.78	1.9	3.63	3.4	6.43	4.9	9.13
0.5	0.97	2.0	3.81	3.5	6.61	5.0	9.30
0.6	1.17	2.1	4.00	3.6	6.79	5.1	9.48
0.7	1.36	2.2	4.19	3.7	6.98	5.2	9.65
0.8	1.55	2.3	4.37	3.8	7.16	5.3	9.83
0.9	1.74	2.4	4.56	3.9	7.34	5.4	10.01
1.0	1.93	2.5	4.75	4.0	7.52	5.5	10.18
1.1	2.12	2.6	4.94	4.1	7.71	5.6	10.36
1.2	2.31	2.7	5.13	4.2	7.89	5.7	10.54
1.3	2.50	2.8	5.32	4.3	8.07	5.8	10.71
1.4	2.69	2.9	5.51	4.4	8.25	5.9	10.89

This completed the theory of the wedge so far as ordinary white light was concerned, and as applicable to the great majority of stars. But as a not inconsiderable number of the latter exhibit colour more or less decided, it seemed desirable to test the wedge in respect of such colours as could be produced from coloured glass and coloured solutions of a definite optical character, in regard to their spectra. The result was that in the case of coloured lights not violently or very strongly pronounced, the wedge was found to be equally absorbent throughout, and the mean wedge interval for one magnitude of such light was found to be the same as for white light. To this point it will be necessary to refer again, in considering the sources of error that might be supposed to effect the results given in this volume.

It should be here noted that a second wedge (B) was examined in the same manner as the former (A). The material and general construction are the same, with the exception that it was made a little steeper for convenience and control over the other. This wedge was found to be practically uniform throughout, and the wedge interval for one magnitude was found to be 0.385 inch.

The Method of using the Wedge Photometer.

About ten stars were selected for a night's work, such that they could be observed at nearly the same altitude as Polaris, in order that it might not be necessary to correct the measures for absorption of light by the atmosphere. The two Photometers A and B were attached to two telescopes, the one of four inches, and the other of three inches aperture, each telescope being in a separate dome; the four-inch telescope being under the charge of Mr. Plummer the Senior Assistant, and the other, under Mr. Jenkins the Junior Assistant: each observer was independently to measure the several stars specified. The complete plan was that Polaris should be extinguished at the beginning, the middle, and the conclusion of the observations. The readings of the wedge were taken five times with the full aperture of the two telescopes. A cap was then placed on each object-glass, reducing the linear aperture to one half, and five readings for extinction were again made. The reason for thus altering the aperture was to establish thereby a check on the former sets of five measures by means of a virtually new instrument.

Each of the other stars whose magnitude was to be compared with that of Polaris was then observed altogether with twenty extinctions, consisting of four independent sets of five extinctions. The wedge readings were usually brought to me on the following morning, and were reduced to magnitude by the method shown in the example on page xxi, on the scale that the magnitude of Polaris should be 2.05. If during the observations any suspicious circumstance had arisen regarding the clearness of the sky in the neighbourhood of Polaris or of the stars observed, the sky was scrutinized out of doors and the observations if necessary discontinued.

The observations of a few nights, which in prudence might properly have been rejected at the time, on account of an unfavourable sky, but which have been inserted in the *Memoirs of the Roy. Astr. Soc.*, have been displaced in the present volume by the substitution of measures, taken under more favourable circumstances: such cases are invariably mentioned in the notes to each constellation.

In order that a correct judgment may be formed of the amount of precision, which may on the average be expected to prevail in the results given in this volume, I now append a tabular exposition of the individual determinations of sixteen stars, each of which has been observed on at least ten separate nights, and most of them with one hundred and twenty extinctions in the aggregate. Each determination is derived from measures made by both observers, with their separate instruments. The result will be found to be, that provided the meteorological circumstances are favourable, little or nothing is gained in point of accuracy over a single night's work of the twenty extinctions, by any multiplication of the measures. From an inspection of the notes attached to the Catalogue, it will be observed that this remark also is fully borne out in the case of a large number of other stars, which have been more or less frequently observed.

TABLE II.

Particulars of numerous repeated measures of sixteen stars.

Date, 1880+	Individual Magnitudes	Number of Extinctions	Date, 1880+	Individual Magnitudes	Number of Extinctions	Date, 1880+	Individual Magnitudes	Number of Extinctions	Date, 1880+	Individual Magnitudes	Number of Extinctions
β Ursæ Majoris.			a Ursæ Majoris.			γ Ursæ Majoris.			δ Ursæ Majoris.		
2.913	2.14	20	2.913	1.90	20	2.938	2.34	20	2.938	3.39	20
3.062	2.20	10	3.062	1.92	10	3.062	2.31	10	3.062	3.37	10
3.070	2.18	10	3.070	1.85	10	3.070	2.27	10	3.070	3.36	10
3.078	2.15	10	3.078	1.90	10	3.095	2.20	10	3.095	3.36	10
3.081	2.09	10	3.081	1.98	10	3.100	2.31	10	3.100	3.53	10
3.089	2.09	10	3.089	1.89	10	5.467	2.45	10	5.467	3.40	10
3.092	2.10	10	3.092	1.92	10	5.483	2.19	10	5.483	3.43	10
5.467	2.19	10	5.467	1.88	10	5.485	2.28	10	5.485	3.55	10
5.483	2.27	10	5.483	1.81	10	5.505	2.47	10	5.505	3.34	10
5.485	2.30	10	5.485	1.80	10	5.514	2.13	10	5.514	3.56	10
						5.519	2.15	10	5.519	3.37	10
						5.533	2.45	10	5.533	3.35	10
Mean	2.17	...		1.89	...		2.30	...	...	3.41	...
ε Ursæ Majoris.			ζ Ursæ Majoris.			η Ursæ Majoris.			β Cassiopeiæ.		
2.938	1.80	20	2.938	1.95	20	2.957	1.75	20	2.835	2.32	20
5.467	1.70	10	5.467	2.08	10	5.467	1.69	10	5.514	2.34	10
5.483	1.91	10	5.483	2.18	10	5.483	1.79	10	5.519	2.40	10
5.485	1.75	10	5.485	2.14	10	5.485	1.83	10	5.533	2.35	10
5.505	1.70	10	5.505	2.16	10	5.505	1.67	10	5.552	2.36	10
5.514	1.79	10	5.514	2.04	10	5.514	1.79	10	5.565	2.52	10
5.519	1.89	10	5.519	2.23	10	5.519	1.88	10	5.571	2.56	10
5.533	1.80	10	5.533	2.02	10	5.533	1.80	10	5.574	2.43	10
5.552	1.85	10	5.552	2.09	10	5.552	1.75	10	5.697	2.42	10
5.565	1.79	10	5.565	2.03	10	5.565	1.73	10	5.724	2.32	10
Mean	1.80	...	...	2.09	...	...	1.77	...	...	2.40	...

TABLE II—*continued.*

Date, 1880+	Individual Magnitudes.	Number of Extinctions.	Date, 1880+	Individual Magnitudes.	Number of Extinctions.	Date, 1880+	Individual Magnitudes.	Number of Extinctions.	Date, 1880+	Individual Magnitudes.	Number of Extinctions.
γ Cassiopeiæ.			δ Cassiopeiæ.			α Cygni.			γ Cygni.		
2.810	2.19	20	2.810	2.89	20	2.430	1.38	20	2.430	2.28	20
5.514	2.27	10	5.514	2.89	10	3.068	1.31	10	3.068	2.23	10
5.519	2.22	10	5.519	2.94	10	3.148	1.33	10	5.533	2.35	10
5.533	2.21	10	5.533	2.93	10	3.164	1.28	10	5.552	2.31	10
5.552	2.42	10	5.552	3.02	10	3.167	1.25	10	5.565	2.17	10
5.565	2.50	10	5.565	2.88	10	3.170	1.34	10	5.574	2.18	10
5.571	2.55	10	5.571	2.91	10	5.533	1.39	10	5.724	2.31	10
5.574	2.34	10	5.574	2.92	10	5.552	1.42	10	5.732	2.18	10
5.724	2.34	10	5.697	2.84	10	5.565	1.25	10	5.738	2.21	10
5.732	2.24	10	5.724	2.81	10	5.574	1.30	10	5.749	2.31	10
5.738	2.14	10	5.732	2.94	10				5.751	2.22	10
5.749	2.27	10									
5.751	2.27	10									
Mean	2.23	...	...	2.91	...	...	1.33	...	...	2.28	...
η Persei.			α Andromedæ.			μ Andromedæ.			R. 204.		
2.699	4.14	20	2.685	2.01	20	2.693	3.98	20	4.012	5.92	20
2.702	4.13	20	2.693	2.05	20	5.692	3.94	10	5.697	5.95	10
2.705	4.09	20	2.697	2.09	20	5.692	3.89	10	5.724	6.11	10
2.708	4.19	20	5.533	2.02	10	5.697	3.90	10	5.724	6.03	10
3.103	4.14	10	5.552	2.24	10	5.724	3.95	10	5.732	6.01	10
3.106	4.05	10	5.565	2.17	10	5.724	3.96	10	5.735	5.92	10
3.108	4.13	10	5.574	2.02	10	5.732	3.85	10	5.738	5.87	10
3.169	4.13	10	5.697	2.00	10	5.735	3.91	10	5.749	5.86	10
3.174	4.09	10	5.721	2.05	10	5.738	3.96	10	5.752	5.88	10
3.207	4.14	10	5.732	2.03	10	5.749	3.91	10	5.757	5.99	10
			5.749	2.02	10	5.752	3.98	10	5.782	5.98	10
			5.751	2.05	10	5.757	3.97	10	5.792	5.97	10
						5.782	3.93	10			
						5.792	3.97	10			
Mean	4.12	...	...	2.05	...	...	3.94	...	...	5.96	...

In explanation of the above table, the case of β Ursæ Majoris may be taken; read in full it is virtually stated in the table that determinations of the magnitude of this star were made on ten nights, extending from November 29, 1882 to June 27, 1885. The magnitude determined on each of these nights is set down in column two, and in column three is printed the number of extinctions on which the several determinations depend. The mean magnitude from all the ten determinations is 2.17 mag., and the number of extinctions one hundred and ten. Similar explanations apply to all the sixteen stars. I may observe in passing that μ Andromedæ and R. 204 Andromedæ were observed for the purpose of watching the variations of the *Nova* which broke out in the Nebula in September 1885.

This table contains many valuable elements for scrutinizing the amount of reliance which may be placed on the Photometric records in this Uranometria. On comparing the results of twenty extinctions on one night with the mean of those obtained from one hundred and ten extinctions on ten nights, it appears that practically nothing has been gained in point of accuracy, as far as these stars are concerned, by the multiplication of the observations.

The precision of the results, both as regards the capacity of the wedges and as evidence of the care bestowed on the observations is, in every respect, satisfactory. It will be noticed, that in the text, the magnitudes do not exactly agree with the mean here set down. The differences are however inappreciable, the reason of any slight disaccordance being, that the majority of these observations were made, after the work had been sent to the press, for the purposes of a stricter scrutiny than had hitherto been applied. I have not selected these sixteen stars from a number of others, but I have given all the records of stars so repeatedly observed. Many other stars also have been frequently measured, but none so often as the above, and all such cases of multiplied observations are given in the ' Notes.'

It will be observed that when ten extinctions have been made, the results are not always quite so closely accordant with the mean as is the case with twenty: but in general, and with few exceptions, the agreements are highly satisfactory. An astronomer will not expect from photometric observations that close accordance which he has a right to demand in bisecting a division of a circle with a microscope, or in the transits of a star across the wires of a meridian circle. It may also properly be remarked that in stellar photometry the observer has no amplification by lenses to assist him, but his attention is directed to the point of disappearance of evanescent lights. Not only so, but the results here given are differential, and dependent upon the errors of *two* fallible measures. The circumstance, which though perhaps unexpected is not less satisfactory, is the practical constancy of the physiological impression on the organs of sight, which is unmistakably manifested by the accordance of delicate measures, taken by two observers, using two instruments, under varying circumstances, and during a considerable lapse of time. On examining the table, it will be found that the mean error of a single *determination* of magnitude, as given by ten or twenty *extinctions*, is under the tenth of a magnitude. The mean error of a single *extinction* is .3 of a magnitude. Of course, this is exclusive of any systematic errors, whether of instrument or observation which *might* exist.

Freedom of the Measurements from the influence of the background.

Experience has led me to the conclusion, that this form of photometer (the wedge) is remarkably free from error in the measurement of light, notwithstanding the illumination of the background. In making this remark, I wish it to be carefully noticed that I am here, and all along, throughout these photometrical researches, confining myself to such discriminations of variation of light as practically fall much within the limit of the unavoidable errors of observation. I regard a variation of one-thirteenth in the magnitude (or about seven per cent.) as the least intensity of light which this instrument is capable of measuring when applied to stellar photometry; and, accordingly, in the remarks that I shall make, I shall regard smaller variations of light than one-thirteenth of a magnitude as practically not influencing the correctness of results; and I know of no measures of stellar light-intensity, excepting those made at Oxford, which reach even this limit of accuracy. I admit that Fechner[1] has referred to experiments on shadows, which show that the eye is, under special circumstances, capable of appreciating an alteration in the illumination of a sheet of paper, arising from the introduction of about one-hundredth of the original light. This experiment has been carefully repeated at the Oxford University Observatory, and Fechner's result has been there verified by observations also in an entirely different form. But it must not be overlooked that in such researches, the comparison is made by means of the juxtaposition of illuminated surfaces of considerable extent. Moreover, the observation is made with the use of both eyes. Independently of this, the method of shadows, apparently so accurate, does not admit of application to the heavens. In Steinheil's photometer, used by Seidel, the comparisons are indeed made by the juxtaposition of stellar discs (or surfaces) not stellar points, as is the case in the Zöllner and in all adaptations of polarization to stellar photometry, but in all these cases, including Steinheil's, the practical limit to the discriminating power of the photometer, is from eight to nine per cent. of the whole light[2].

It is very necessary to bear these limitations in mind, in forming a judgment of the reliability of the wedge photometer, when viewing stars on faintly illuminated backgrounds, such as those which occur in practice. This faint illumination arises from one of two causes, generically different in character. The first is produced by the diffused or scattered light in the atmosphere or sky; the second arises from the presence of other stars in the field of the telescope as limited in the photometer. Each of these must be considered separately from the other.

When stars are viewed on moonlight nights with the wedge photometer, it might probably be surmised (though not after due consideration) that an effect would be produced on the measure of the light of a star owing to this background. But, in the first place, it would be injudicious to measure a star very close to the moon itself; and in practice it would be avoided. In the second place, it has been shown[3] practically, from the examination of some five thousand extinctions of Polaris, made during

---

[1] Leipzig, Abhandlungen der Mathematisch-Physischen Classe, p. 467.

[2] See Wilsing, Astron. Nachrichten, vol. cix, p. 50, where, from experiment, he assigns a greater probable error to the Zöllner photometer than to the wedge.

[3] Monthly Notices, vol. xlv, p. 412.

all phases of the Moon, that there is not the slightest trace observable of effects due to moonlight or its absence. Nor indeed could it be otherwise; for the dispersed or scattered light of the atmosphere or of a blue sky, is not, as in the case of stars, concentrated at the focus of the object-glass, but remains very nearly as it would, if viewed with the tube alone from which the object-glass and eye-piece are removed. Practically this scattered light is, on a moonlight night, extinguished at one inch from the thinnest end of the wedge, and long before a sixth magnitude star, or the faintest in this catalogue, approaches extinction.

The second case contemplated is that in which other stars are present in the same field of view with the star which is the subject of measurement. The photometric field of view in the four-inch telescope is confined to a circle of 0.02 inches diameter, equivalent to about 80″. A reference to Argelander's charts will indicate the fact, that in very few instances indeed will so contracted a field, comprise, together with a sixth magnitude star, any star ranging from the seventh to the tenth magnitude. A seventh magnitude is very rare: if it be within (say) 20″ of the star to be measured, then it would be placed in the category of double stars, and its character would be referred to in the Notes. It would, if viewed as one point of light, coalesce with the sixth magnitude, and the combined light would be of 5.6 magnitude. If, on the other hand, the small star, or the two or three small stars in the field be detached from the brighter star, they would produce distinct impressions on the retina, and would be completely extinguished by the wedge long before the extinction of the brighter, and could produce absolutely no influence on its measure in the photometer. Precisely the same sort of reasoning applies to the measurement of fainter stars in a larger telescope, say of twelve inches or any greater aperture. The photometric field in such telescopes is still more restricted, and the very faint stars possibly accompanying the one to be measured, would either coalesce as a double or multiple star, and be noted as such, or they would be extinguished by the wedge long before the extinction of the brighter.

The effect of the coalescence of two stars upon the magnitude of the brighter, is shown in the following short table:—

Difference in magnitude of two stars whose lights coalesce.	Resulting addition of brightness to the brighter star in magnitude.
0.0	0.75
0.5	0.53
1.0	0.37
1.5	0.25
2.0	0.16
3.0	0.07
4.0	0.03
5.0	0.01

These considerations are sufficient to indicate the practical freedom of this photometer from the influence of the greater or less density of stars illuminating various regions of the heavens (always within the limits of accuracy prescribed).

So far, the above conclusions are drawn chiefly from abstract grounds. Practically, a comparison of the photometric magnitudes of a large number of stars between the fourth and six-and-a-half magnitudes has been made with those given in the Harvard Photometry. Not a trace of any systematic difference has been found in the magnitudes in the two catalogues, so far as is due to the varying stellar densities of the regions where these comparisons have been made [1].

Another cause of conceivable error, and one to be guarded against in the use of the photometer, arises from the possible fatigue of the eye during any considerable series of measures on any single night, whereby the want of constancy in the appreciation of faint lights, might operate unfavourably. In order to assure myself of the non-existence of this sort of error, a large number of extinctions of Polaris made on separate nights, have been examined. These extinctions of Polaris are, as has been explained, made three times on each night, viz. at the beginning, the middle, and the end of the series. No trace of a systematic difference is found indicating symptoms of varying sensitiveness of the observer's eye, depending on these sequences of time. Moreover, as the measures are strictly differential, and as the mean of the three sets of readings of the extinction of Polaris (which rarely differ much *inter se*), is used for all the observations of the series, the mean condition of the eye is represented throughout. In point of fact, relief is afforded by continual breaks in the work of actual observation, connected with the shifting of the domes, with the identification of the stars, and with the recording the results; the whole series also seldom occupies more than three hours in the night. All these circumstances combine to prevent any undue strain on the eye of the observer, during the limited time of his work.

The action of this form of photometer in relation to colour, seems to be in pretty much the same category as any other photometer, not excepting even Zöllner's ingenious contrivance. The colours of the stars, with a few well-known exceptions, are faintly marked. The action of the wedge on *such* lights was carefully examined by means of coloured solutions and coloured glasses, and by the spectroscope, as is detailed in the *Memoirs of the Royal Astronomical Society*, vol. xlvii, pp. 395–6, and, as has been already mentioned at p. xiii, no defect was discovered within the prescribed limits of error of observations. The material of the wedge has a selective absorption for the red part of the spectrum near to the line *B*, and every case of a star reported to have a red tinge, is referred to in the notes appended to each constellation. It is, however, very remarkable that out of all these numerous instances of stars of a more or less reddish tint, there is very rarely any difference between the Oxford and the

---

[1] Monthly Notices, vol. xlv, p. 411.

Harvard magnitude, greater than the mean difference for all the stars examined in both catalogues. The question of colour seems to be, to a very considerable extent, subjective and physiological, and, at present, out of the domain of exact science. Possibly the arrangements described by the present Astronomer Royal in the Monthly Notices, January 1874, may be developed with success in this direction.

Example of Reductions.

In the following table are exhibited the measures and processes adopted for the final determination of the magnitude of a star: here 72 Tauri.

TABLE III.

Original Observations of the difference of Magnitude between 72 Tauri and Polaris.

WEDGE A.

Date and Instrument.	Original Wedge Reading of Extinction of 72 Tauri.	Mean Wedge Readings of Extinction of Polaris at beginning, middle, and end of series.	Difference or Wedge Interval in inches.	Equivalent of Wedge Interval in Magnitude. *See Table I.*	Assumed Magnitude of Polaris.	Resulting Magnitude of 72 Tauri.	Finally Adopted Magnitude.
	in.	in.					
1884, Dec. 20.	2.65	4.602					
Wedge A	.53	4.592					
4-inch Aperture	.72	4.606					
	.65						
	.71						
Mean	2.652	4.600	1.948	3.56	2.05	5.61	
Wedge A	1.90	3.828					
2-inch Aperture	.83	3.868					
	.79	3.840					
	.92						
	.86						
Mean	1.862	3.845	1.983	3.69	2.05	5.74	5.65

TABLE III—*continued.*

WEDGE B.

Date and Instrument.	Original Wedge Reading of Extinction of 72 Tauri.	Mean Wedge Readings of Extinction of Polaris at beginning, middle, and end of series.	Difference or Wedge Interval in inches.	Equivalent of Wedge Interval in Magnitude. 1 mag. = .385 in.	Assumed Magnitude of Polaris.	Resulting Magnitude of 72 Tauri.	Finally Adopted Magnitude. Mean of the four.
	in.	in.					
1884, Dec. 20.	1.30	2.682					5.65
Wedge B	.27	2.698					
3-inch Aperture	.29	2.682					
	.34						
	.26						
Mean	1.292	2.687	1.395	3.62	2.05	5.67	
Wedge B	0.71	2.076					
1½ inch Aperture	.60	2.084					
	.74	2.062					
	.70						
	.79						
Mean	0.708	2.074	1.366	3.54	2.05	5.59	

The first column contains the date: the second is the record of the wedge readings for each extinction of the star in the wedge A, with aperture four inches, the mean reading of the five extinctions is 2.652 in. The second column contains the three readings of the extinction of Polaris at the beginning, the middle, and close of the observations. Their mean is 4.600 inches. The equivalent magnitude for 2.652 in Table I is 5.04, and the magnitude for 4.600 in. is 8.60. Finally 8.60 − 5.04 = 3.56. This is the difference of magnitudes of the star and Polaris; and as the latter is conventionally 2.05 mag. the resulting magnitude of the star is, so far as this single measure is concerned, 5.61. Similar wedge readings are then taken with wedge A, after the linear aperture of the telescope to which it is attached is halved. These measures, the mean of which is 1.862, are then combined with the mean of the three wedge readings for Polaris with this aperture, viz. 3.845. Resort is then had to Table I, as before, for these two wedge readings, whence the equivalent difference of magnitude of the star and Polaris is derived, viz. 3.69, and the resulting magnitude of the star is 5.74. In the remainder of the Table are given the details of the extinctions with the wedge B, with apertures of 3 inches and 1½ inches. The resulting magnitudes of the star are 5.67 and 5.59. The final and tabular determination of the magnitude of 72 Tauri is 5.65. The mean deviation of the four individual measures is .05 magnitude; and if on several other nights the determinations of the star's magnitude had been taken, there is little doubt, judging by experience (see page xv), that the magnitude (5.65) would not differ from the general mean of all by one-tenth of a magnitude.

Observations taken at Cairo.

It necessarily formed part of the scheme of photometric work at Oxford to determine the amount of light absorbed by the earth's atmosphere when a star is observed at any varying altitude. Without such determination, the research would have been incomplete, although, so far as the stars in this Uranometria are concerned, a knowledge of the amount of this absorption is rarely required, inasmuch as the stars are in general observed at an altitude so near that of Polaris, that no correction is required for absorption in order to obtain the zenithal magnitude of the star, because the zenithal magnitude of Polaris is assumed at 2.05, and absorption would act equally on the two if observed on the same night and under approximately the same atmospheric circumstances. The detail of the work for the determination of the atmospheric absorption constant at Oxford is all given in *Memoirs of the Roy. Astr. Soc.*, vol. xlvii: the result being to show the necessity of a correction to magnitude, of 0.253 sect. z. I was not wholly satisfied with this determination, excepting so far as it accurately represented the local atmospheric absorption at Oxford, and I thought it desirable to measure it again, in some other locality where the climate is more equable, and the atmospheric circumstance in general more favourable, and more accordant measures might be anticipated. I was also desirous of ascertaining what improvements might be effected in the determination of magnitude in that climate.

Accordingly, I determined on a voyage to Cairo, taking thither Mr. Jenkins, with the telescope and wedge A, hitherto used by Mr. Plummer. I had carefully ascertained before making the exchange, that it was indifferent which wedge was used by either of the two observers, notwithstanding his habituation to the use of one particular instrument. In due time the telescope and Mr. Jenkins were duly installed in the admirable observatory, belonging to the Khedive at Abbasseeyeh, within a stone's cast of the English barracks in the desert. The atmosphere and climate proved to be all that had been stated in their favour, and about three times as much work was there effected in six weeks as was done, or in fact could be done at Oxford: for it was a settled part of the plan, that Mr. Plummer in Oxford was to make the same observations as Mr. Jenkins at Cairo. The results of the expedition are all embodied in the measures recorded in this volume, and greatly add to the confidence I feel in the general accuracy of the work herein exhibited. The atmospheric absorption constant at Cairo proved, as might be expected, slightly less than that at Oxford, but not so much less as I had anticipated. The climate during the time I was in Egypt, February and March, was indescribably agreeable, and equable to a degree unknown in England; but its increased transparency was only such as to admit of the partial visibility of stars exceeding those visible at Oxford by about a fifth of a magnitude. Of course this addition to the number of stars visible would be counted by the thousand rather than by the hundred; nevertheless the gain in transparency alone was in a telescopic sense comparatively trifling;—in steadiness the advantage was enormous during the time mentioned. Probably at other periods it is much less so, but I speak here from conjecture.

The resulting values of atmospheric absorption expressed in magnitude by various observers at various localities are—

		Mag.
Bouguer in Brittany	. .	0.225
Seidel at Munich	. . .	.250
Langley on Mount Etna	. .	.126
Pritchard at Oxford	. .	.253
Pritchard at Cairo	. . . .	.187
Müller at Potsdam	. . . .	.209
Pickering at Harvard	. . . .	.250

I will only add that in the climate of Oxford and probably that of the greater part of Europe the application of this formula is not safe beyond 60° or 65° from the zenith. In a climate such as Cairo it may probably be trusted to 75° or 80° Z. D. But I have restricted the catalogue of this Uranometria to such stars as it is possible to observe at a moderate distance from the zenith or at the same altitude nearly as Polaris, and in this way the difficulty is evaded. Generally speaking, when great accuracy of result is needed it will be found necessary to restrict photometry to observatories favourably situated for the purpose.

The result of the many fairly accordant determinations of the general atmospheric absorption given above, obtained as they are by methods widely different, leave no doubt of their approximate and practical accuracy, with this reservation, that it may be admitted, they are not to be trusted as representing facts anywhere near to the horizon. This remark may go far to explain the doubts expressed by Professor Langley as to the general untrustworthiness of the whole investigation from a theoretical point of view.

Explanation of the Catalogue.

We are now arrived at the description of the Catalogue itself. The first measure was made on Dec. 22, 1871; the last on Oct. 16, 1885. The stars selected are strictly those in Argelander's *Uranometria Nova*, though but few were measured in Oxford, below 10° south of the Equator, on account of the uncertainty of the absorption correction for low altitudes, and of these few, the adopted magnitudes are given in the Notes : not that the *mean* value of this correction is so uncertain, but because my experience is, that it varies very sensibly from night to night. Where my work ends it may be more properly continued at some Southern Observatory, and I trust I may be permitted to hope, that as the precision of modern astronomy advances, each observatory will confine its attention to that work alone, for which its position and equipment especially fit it. With regard to celestial photometry in Oxford, I should have preferred not to carry the observations south of the Equator, but it seemed desirable to extend them to some 10° South, in order to provide an overlap for connecting the Oxford results with those obtainable in some observatory in Southern Latitude.

I have adopted the arrangement of the stars in constellations, mainly because the stars concerned are visible to the naked eye, and are more easily cognizable than if the usual celestial co-ordinates alone were given. The abbreviations made use of in the column of ' Star's designation,' and elsewhere in the volume, are such as are generally recognised by astronomers and consequently need no explanation.

I have given the estimated magnitudes of the *Uranometria Nova*, rather than those in the *Durchmusterung*, not only on account of the care which Argelander bestowed on this special subject, but because the stars in the larger catalogue were all observed telescopically, and therefore are derived from the impressions of memory alone : nevertheless, I believe, the latter represents faithfully the heavens as they are seen. The average deviation from the mean of the usual twenty extinctions is given, rather than the 'probable error,' because the latter term is liable to misinterpretation : moreover, it must not be forgotten that this 'average deviation' is a test, rather of the general accordance of the individual extinctions with themselves, than of the actual accuracy of the final determination. This accuracy may be better estimated from the inspection of the numerous measures on different nights of the stars given on page xv of this Preface.

I have confined the Notes to calling attention to such features of the star in question, which I considered worthy of notice, or likely to affect the determination of magnitude. The spectra of stars are probably of far greater importance and possess more significance than any indication of colour, which not without some care is detectable by the naked eye : consequently whenever there is anything remarkable in these spectra, I have directed attention to the fact so far as the valuable observations of Dr. Vogel, as given in the *Publicationen des Astrophysikalischen Observatoriums zu Potsdam*, Nr. 11, are concerned. Vogel's classes, described in his own words, are—

Classe I.

Spectra, in welchen die Metalllinien nur äusserst zart auftreten oder gar nicht zu erkennen sind und die brechbareren Theile des Spectrums, Blau und Violett, durch ihre Intensität besonders auffallen.

(a) Spectra, in denen, ausser den sehr schwachen Metalllinien, die Wasserstofflinien sichtbar sind und sich durch ihre Breite und Dunkelheit auszeichnen.

(b) Spectra, in denen entweder einzelne Metalllinien nur ganz schwach angedeutet, oder gar nicht zu erkennen sind und die Wasserstofflinien fehlen.

Classe II.

Spectra, in denen die Metalllinien sehr deutlich auftreten. Die brechbareren Theile des Spectrums sind im Vergleich zur vorigen Classe matter. In den weniger brechbaren Theilen des Spectrums treten zuweilen schwache Banden auf.

(a) Spectra mit sehr zahlreichen Metalllinien, die besonders im Gelb und Grün durch ihre Intensität leicht kenntlich werden. Die Wasserstofflinien sind meist kräftig, aber nie so auffallend verbreitert als bei Classe Ia. Bei einigen Sternen sind die Wasserstofflinien schwach, und bei solchen sind dann gewöhnlich in den weniger brechbaren Theilen schwache Banden zu erkennen.

Classe III.

Spectra, in denen ausser dunklen Linien noch zahlreiche dunkel Banden in allen Theilen des Spectrums auftreten und die brechbareren Theile des Spectrums auffallend schwach sind.

(a) Ausser dunklen Linien sind in den Spectren Banden zu erkennen, von denen die auffallendsten nach Violett dunkel und scharf begrenzt, nach Roth matt und verwaschen erscheinen.

(b) Spectra, in denen dunkel, sehr breite Banden zu erkennen sind, von denen die am stärksten hervortretenden nach Roth scharf begrenzt und am dunkelsten sind, nach Violett allmählich erblassen. Sie erscheinen also nach der entgegengesetzten Seite verwaschen, wie die Banden in den Spectren der vorigen Unterabtheilung, auch ist ihre Anzahl geringer als dort. Das Blau und Violett ist gewöhnlich überaus schwach.

Es haben besonders schöne Exemplare der betreffenden Classe und interessante Spectra ! ! ! erhalten. Schön ausgeprägte Exemplare erhielten ! !, deutlich, auf den ersten Blick zu erkennende Spectra endlich haben ! erhalten.

Inasmuch as the photometry of stars exhibiting any salient colour may be open to some doubt, I have noted all such cases, including all Birmingham's stars, though the colours of many are often practically not recognizable.

When the measures in this Uranometria were about three-fourths completed, the 'Harvard Photometry' was published, and inasmuch as this work is one of unquestionable importance, it became desirable to compare the results there given, with my own. Wherever there was a difference exceeding a third of a magnitude between the two catalogues, the stars were re-measured at Oxford. In the great majority of instances, the previously existing measures were re-obtained. Still, it may not be assumed that my determinations are always free from error, and I have accordingly given the Harvard magnitude, as well as the Oxford, but as all the stars in both catalogues are compared with Polaris, and as in the American Catalogue 2.15 is assumed as its conventional magnitude, whereas 2.05 is assumed in this Uranometria, it is necessary to subtract one-tenth of a magnitude from all those to which H.P. is affixed in the Notes. I did not consider myself warranted in taking the liberty of making the alteration in a more formal manner.

Until Sir John Herschel's photometric researches at the Cape had made it possible to discriminate between the lustres of the brightest stars, these were all grouped together as stars of the first magnitude, though in point of fact they differed from each other enormously in respect of intensity of light. When this fact became apparent a new nomenclature regarding such stars became necessary.

The one adopted, that of the magnitude *nothing*, and still more the magnitude *minus*, necessarily leads to confusion. I have therefore ventured to propose another designation, to the effect that the symbol + placed before a magnitude shall intimate that its lustre expressed numerically is so many magnitudes or portions of a magnitude brighter than the first; so that + 0.4 and + 1.3 shall indicate respectively, four-tenths of a magnitude, and one and three-tenths of a magnitude, brighter than the first. In this way the magnitude of Sirius is designated by + 1.95, indicating that it is very nearly two magnitudes brighter than a star, such as α Aquilæ or Spica, which stars are very properly regarded as approximately of the first magnitude on Pogson's scale. In connection with this I append the photometric magnitudes of such stars as have heretofore been grouped together, as of the first order of lustre.

Star's Name.	Magnitude.	Star's Name.	Magnitude.
Sirius	+ 1.95	Spica	+ 0.04
Rigel	+ 1.03	α Orionis	+ 0.02
Capella	+ 0.92	α Aquilæ	1.04
Vega..............	+ 0.86	Aldebaran	1.12
Arcturus	+ 0.69	Antares	1.13
Procyon	+ 0.50	Pollux	1.36

As a matter of fact, and not without some interest, it may be stated that, on comparing the results of these photometric measures with the telescope estimates of the same stars in the *Durchmusterung*, the mean difference is found to be .06 in the magnitude of a single star. The two catalogues, therefore, notwithstanding many individual discordances, may in the average, and on the whole, be regarded as in substantial agreement.

The researches in this volume, being the first that have been completed on the instrumental photometry of the stars by an European astronomer, must have historical relations. If, on examination, they shall also be found to possess, as I confidently expect they will, an appreciable scientific value, I desire to record that much of it is due to the skilful and unremitting attention of my two assistants Mr. W. Plummer, F.R.A.S. and Mr. C. Jenkins, F.R.A.S., by whom the necessary measurements, exceeding seventy thousand, were made.

ANDROMEDA.

Reference Number.	Star's Designation.	R.A. 1890.	N.P.D. 1890.	Adopted Zenithal Magnitude. Polaris 2.05.	Average Deviation in Magnitude	Date, 1880 +	Mag. Argel. Uran.
		h. m.	° ′				
1	o	22 56.9	48 16	3.74	0·06	2.677	4...3
2	2	22 57.5	47 50	5.19	.05	3.960	6
3	3	22 59.2	40 33	4.93	.06	3.960	5...6
4	4	23 2.6	44 13	5.42	.05	3.960	6
5	7	23 7.5	41 12	4.92	.04	4.012	5
6	Bradley 3084......	23 11.7	37 23	5.68	0.06	3.960	6
7	B.A.C. 8110	23 12.1	45 26	6.41	.20	3.973	6
8	8	23 12.6	41 35	5.02	.02	4.012	5...6
9	9	23 13.2	48 50	6.22	.09	3.973	6
10	11	23 14.4	41 59	5.53	.05	4.012	6
11	12	23 15.6	52 25	5.96	0.09	4.032	6
12	13	23 21.8	47 42	5.89	.02	3.973	6
13	14	23 25.9	51 22	5.44	.09	4.012	6
14	15	23 29.3	50 22	5.51	.05	4.012	6
15	λ	23 32.2	44 8	3.69	.06	...	4
16	ι	23 32.7	47 20	4.56	0.05	...	4
17	18	23 33.8	40 8	5.41	.05	4.012	6...5
18	κ	23 35.0	46 16	4.34	.06	2.677	4
19	ψ	23 40.6	44 11	5.04	.04	3.973	5
20	L. 46676	23 44.1	54 11	6.02	.02	4.032	6
21	R. 6226............	23 51.5	47 57	6.25	0.04	3.973	6
22	W.B. 23ʰ-1073 ..	23 53.2	58 15	6.45	.00	4.032	6
23	R. 6265............	23 56.1	48 16	6.19	.09	4.012	6
24	α	0 2.7	61 31	2.05	.05	2.685	2
25	22	0 4.6	44 32	5.05	.06	3.973	5 ..6

No. 6. B.A.C. assigns it to Cassiopeia.
No. 8. Called by Arg. red : the colour is not salient.
No. 15. 3.69 is the mean of two measures (3.64, 3.74' observed with twenty extinctions 1882.677 and with ten, 1885.102. Schmidt calls this star golden yellow.
No. 16. 4.56 is the mean of two measures (4.58, 4.54' observed with twenty extinctions 1882.677 and with six, 1885.102. H.P.=4.30.
No. 24. Called δ Pegasi in the Almagest.
No. 25. Σ 3, Dist. 5″. The magnitude is that of the combination.

Reference Number.	Star's Designation.	R.A. 1890.	N.P.D. 1890.	Adopted Zenithal Magnitude. Polaris 2.05.	Average Deviation in Magnitude.	Date, 1880+	Mag. Argel. Uran.
		h. m.	° '				
26	W.B. 0ʰ-152......	0 7.7	63 38	6.24	0·06	4.032	6
27	23	0 7.8	49 34	5.89	.09	3.973	6
28	W.B. 0ʰ-181......	0 8.4	57 24	5.96	.06	4.032	6
29	W.B. 0ʰ-210......	0 9.5	63 20	6.00	.04	4.032	6
30	θ	0 11.4	51 56	4.45	.03	4.012	5...4
31	σ	0 12.6	53 49	4.39	0.02	...	4...5
32	P. 0ʰ-38	0 12.9	59 5	5.75	.06	4.032	6
33	W.B. 0ʰ 339 1st star	0 14.7	59 40	5.88	.09	4.032	⎫
34	W.B. 0ʰ-339 2nd star	0 14.8	59 39	6.95	.01	4.032	⎬ 6
35	L. 367 1st star ...	0 15.0	57 42	6.03	.05	4.029	6
36	L. 367 2nd star...	0 15.2	57 38	6.91	0.03	4.029	6
37	ρ...............	0 15.3	52 38	5.44	.04	4.029	6
38	B.A.C. 100	0 22.3	46 13	5.44	.02	4.012	6
39	28	0 24.3	60 51	5.35	.04	4.012	6.. 5
40	P. 0ʰ-122	0 30.6	63 21	6.32	.04	4.012	6
41	B.A.C. 152	0 30.8	46 7	5.38	0.10	4.012	6
42	π	0 31.0	56 53	4.24	.05	...	4
43	B.A.C. 158	0 31.5	55 12	5.79	.07	4.012	6
44	ε...............	0 32.8	61 17	4.29	.11	2.685	4
45	δ	0 33.4	59 44	3.18	.11	...	5...4
46	32	0 35.2	51 9	5.49	0·04	4.032	5
47	R. 204	0 40.1	45 45	5.92	·06	4.012	6
48	ζ...............	0 41.5	66 20	4.13	·04	2.693	4
49	ν...............	0 43.8	49 31	4.62	·03	2.693	4...5
50	R. 225	0 44.1	45 36	5.65	·09	4.040	6
51	μ	0 50.6	52 6	3.98	0·05	2.693	4
52	η	0 51.3	67 10	4.41	·11	2.677	5
53	P. 0ʰ-242	0 52.2	56 38	6.48	·06	4.012	6
54	R. 288	0 53.8	45 53	5.89	·05	4.040	6
55	39	0 56.7	49 15	6.07	·06	3.968	6

No. 31. **4.39** is the mean of two measures (4.37, 4.40) observed with twenty extinctions 1882.667, and with six, 1885.102.

No. 42. The observation of 1882.685 (See Mem. R.A.S. vol. 47, p. 420) is rejected, the night being hazy. There is a faint companion 36" distant not observed.

No. 45. Birm., No. 8. Noted by Schmidt as of a deep golden yellow. **3.18** is the mean of two measures (3.23, 3.14) observed 1882.682 and 1882.693, each involving twenty extinctions.

No. 54. Σ 79, Dist. 8". Observed as one mass.

Reference Number.	Star's Designation.	R.A. 1890.		N.P.D. 1890.		Adopted Zenithal Magnitude. Polaris 2.05.	Average Deviation in Magnitude.	Date, 1880+	Mag. Argel. Uran.
		h.	m.	°	′				
56	41	1	1.7	46	39	5.17	0.03	3.968	5
57	φ	1	3.1	43	21	4.38	.02	2.693	4...5
58	β	1	3.6	54	58	2.21	·05	2.688	2...3
59	44	1	4.1	48	30	5.83	.03	4.012	6
60	45	1	4.9	52	52	5.93	.09	4.051	6
61	ξ	1	15.9	45	3	5.18	0·09	4.040	5
62	47	1	17.4	52	51	5.75	.06	4.051	6
63	P. i. 69	1	19.8	47	7	5.99	.04	4.040	6
64	W.B. 0ʰ-378.....	1	19.9	55	59	6.51	.10	4.051	6
65	ω	1	21.0	45	10	4.90	.02	4.018	5
66	49	1	23.5	43	34	5.38	0·05	4.018	6
67	W.B. 0ᵇ-530......	1	26.5	54	43	6.59	.09	4.051	6
68	P. i. 104.........	1	27.9	53	20	6.10	.06	4.051	6
69	υ	1	30.3	49	8	4.26	.04	2.693	4...5
70	χ	1	32.8	46	10	5.18	.07	4.018	5 ..6
71	B.A.C. 501	1	34.0	47	15	5.78	0.02	4.018	6
72	τ	1	34.1	49	59	5.01	.07	4.051	5
73	P. i. 142.........	1	35.0	47	56	5.41	.09	4.018	6
74	55	1	46.7	49	49	5.69	.05	4.018	6
75	B.A.C. 579	1	49.4	53	16	5.79	.06	4.075	} 5
76	56	1	49.4	53	16	5.84	0.03	4.075	
77	γ¹	1	57.1	48	12	2.14	.03	2.693	} 2...3
78	γ²	1	57.1	48	12	4.86	.04	2.693	
79	58	2	1.8	52	40	5.08	.02	...	5
80	59 1st star........	2	4.2	51	29	6.33	.04	4.075	} 6
81	59 2nd star	2	4.2	51	29	6.12	0·04	4.075	
82	60	2	6.3	46	17	5.08	.04	4.018	5...6
83	62	2	12.2	43	8	5.40	.12	4.018	5...6
84	63	2	13.7	40	21	5.64	.04	4.018	6
85	B.A.C. 727	2	16.0	49	6	6.07	.03	4.018	6
86	64	2	17.3	40	30	5.44	0.05	4.018	6
87	65	2	18.5	40	13	5.08	.09	4.018	5

No. 57. OΣ 515. The magnitude is that of the combination.

No. 58. Birm., No. 17. Called red by Schmidt. The red colour is not salient.

No. 61. Variability suggested by Pigott (see Phil. Trans. 1786, p. 203).

No. 69. The Greek letter, which is in Bayer's Map (1603), is omitted in the B.A.C.

Nos. 75, 76. Σ 4¹, Distance 177″.

No. 77. Σ 305, Dist. 10″. The colour is intensely yellow.

No. 78. The magnitude is that of the combination, Dist. <1″. The combined colour is green.

No. 79. 5.08 is the mean of two measures (5.06, 50.9) observed 1884.075 with twenty extinctions and with six, 1885.102. Nos. 80, 81. These stars form Σ 222, Dist. 16″.

No. 82. Birm., No. 37. Pale orange. No. 87. Birm., Second list of Addenda. Red.

AQUARIUS.

Reference Number.	Star's Designation.	R.A. 1890.	N.P.D. 1890.	Adopted Zenithal Magnitude. Polaris 2.05.	Average Deviation in Magnitude.	Date, 1880 +	Mag. Argel. Uran.
		h.　m.	o　′				
1	1.................	20 33.8	89 54	5.50	0.03	4.598	5
2	3.................	20 41.9	95 26	4.84	.03	4.598	4...5
3	5.................	20 46.3	95 55	5.74	.08	4.764	5
4	μ	20 46.7	99 24	4.88	.06	4.598	5...4
5	P. xx. 360	20 48.1	97 18	6.51	.06	4.771	6
6	11	20 54.8	95 9	6.50	0.07	4.764	6
7	12	20 58.3	96 15	5.69	.08	4.771	5...6
8	15	21 12.4	94 59	6.04	.06	4.771	6
9	16	21 15.3	95 2	6.12	.06	4.764	6
10	21	21 19.6	94 2	5.84	.05	4.771	6
11	β	21 25.8	96 3	3.08	0.03	4.771	3
12	P. xxi. 190.........	21 29.6	94 28	5.91	.04	4.712	6
13	ξ	21 31.9	98 21	4.71	.06	4.764	5...4
14	Σ 2809	21 31.9	90 53	6.02	.06	4.731	6
15	25 (d)	21 34.0	88 15	5.35	.06	4.719	6...5
16	26	21 36.6	89 13	5.99	0.09	4.771	6
17	P. xxi. 320.........	21 48.4	94 48	6.01	.06	4.764	6
18	Σ 2838	21 48.8	93 50	6.48	.07	4.712	6
19	P. xxi. 345.........	21 52.5	95 57	6.41	.05	4.739	6
20	28	21 55.4	89 56	5.85	.05	4.731	6
21	30	21 57.5	97 3	5.65	0.06	4.771	5...6
22	6.................	21 57.6	92 41	4.60	.07	4.712	5...4
23	32	21 59.1	91 26	5.60	.03	...	6
24	a	22 0.1	90 51	3.04	.07	4.731	3
25	P. xxi. 421.........	22 4.6	94 26	6.07	.07	...	6

No. 7. Σ 2745, Dist. 3″. Observed as one star.
No. 11. This star is slightly red or orange.
No. 14. Dist. 31″. The larger star observed.
No. 15. = 6 Pegasi. See Introduction to B.A.C., p. 75.
No. 17. Variability suspected by Gore.
No. 18. Σ 2838, Dist. 22″. The brighter star observed.
No. 23. 5.60 is the mean of three very accordant observations made in 1884.764, and 1885.464 and .467. H.P. = 5.24.
No. 25. 6.07 is the mean of three accordant observations made in 1884.739, and 1885.464 and .467. H.P. = 5.85.

Reference Number.	Star's Designation.	R.A. 1890.	N.P.D. 1890.	Adopted Zenithal Magnitude. Polaris 2.05.	Average Deviation in Magnitude.	Date. 1880+	Mag. Argel. Uran.
		h. m.	° ′				
26	P. xxii. 17	22 7.0	95 16	6.51	0.05	4.771	6
27	θ	22 11.0	98 20	4.30	.06	4.731	4...5
28	44	22 11.4	95 56	5.88	.08	4.719	6...5
29	ρ...................	22 14.4	98 22	5.50	.04	4.739	5...6
30	γ	22 15.9	91 56	3.97	.05	4.712	4...3
31	51	22 18.4	95 24	5.04	0.06	4.771	6
32	π	22 19.7	89 11	4.84	.05	4.764	5...4
33	ζ¹	22 23.2	90 35	3.71	.05	4.712	3...4
34	L. 43974	22 25.5	97 6	6.43	.04	4.764	6
35	60	22 28.4	92 8	6.14	.06	4.719	6
36	η	22 29.7	90 41	4.12	0.05	4.739	4...3
37	κ...................	22 32.1	94 48	5.43	.03	4.771	5
38	67	22 37.5	97 32	6.45	.07	...	6
39	λ	22 46.9	98 10	3.82	.07	4.764	4
40	P. xxii. 250	22 49.5	95 34	6.05	.03	...	6
41	83	22 59.4	98 17	5.61	0.08	4.739	6...5
42	φ	23 8.6	96 38	4.32	.06	4.764	4...5
43	L. 45521	23 9.8	94 5	5.51	.06	4.731	5...6
44	χ	23 11.1	98 20	5.52	.06	...	5...6
45	96	23 13.7	95 43	5.76	.04	4.739	6...5
46	P. xxiii. 96.........	23 23.8	95 8	6.09	0.08	4.764	6...5

No. 33. Σ 2909, Binary. Period uncertain. The observation is of the combination.
No. 38. 6.45 is the mean of three accordant measures, made on 1884.764, and 1885.464 and .467. H.P. = 6.20.
No. 39. Birm., No. 621. The red colour is not salient.
No. 40. 6.05 is the mean of three accordant measures made on 1884.719, and 1885.464 and .467. H.P. = 5.77.
No. 42. Birm., No. 633. The red colour is noticeable.
No. 44. Birm., No. 634. The red colour is not salient. 5.52 is the mean of three accordant measures made on 1884.771, and 1885.464 and .467. H.P. = 5.15.
No. 46. The B.A.C. assigns this star to Pisces.

AQUILA.

Refer-ence Number.	Star's Designation.	R.A. 1890.	N.P.D. 1890.	Adopted Zenithal Magnitude. Polaris 2.05.	Average Deviation in Magnitude.	Date, 1880+	Mag. Argel. Urau.
		h. m.	° ′				
1	4	18 39.3	88 3	5.03	0·04	3.519	5
2	5	18 40.7	91 5	6.09	.05	...	6
3	L. 35150	·18 47.0	76 10	6.00	.08	3.519	6
4	L. 35281	18 50.7	91 56	6.05	.08	3.519	6
5	L. 35421	18 53.4	72 47	5.03	.04	3.604	5
6	11	18 54.0	76 31	5.25	0.04	3.604	5
7	ε	18 54.6	75 5	3.86	.03	2.583	4
8	12	18 55.8	95 54	4.01	.06	2.583	5···4
9	14	18 57.1	93 51	5.74	.08	3.519	6
10	L. 35598	18 58.0	88 21	5.97	.02	3.604	6
11	15	18 59.2	94 12	5.67	0.09	3.519	6
12	ζ	19 0.4	76 18	3.08	.05	2.586	3
13	λ	19 0.4	95 3	3.27	.04	2.586	3···4
14	18	19 1.8	79 6	5.00	.12	3.604	5
15	L. 35851	19 3.0	73 18	5.96	.03	3.604	6
16	19	19 3.6	84 6	5.44	0.04	3.519	5···6
17	20	19 6.7	98 7	5.42	.08	3.519	6
18	21	19 8.2	87 54	5.16	.05	3.527	6···5
19	L. 36207	19 10.3	75 6	5.72	.02	3.604	6···5
20	22	19 11.1	85 22	5.29	.11	3.527	6
21	L. 36268	19 11.4	75 39	5.42	0.02	3.604	6···5
22	L. 36285	19 12.3	88 10	6.19	.07	3.527	6
23	ω	19 12.7	78 36	5.35	.02	3.554	6···5
24	23	19 12.9	89 7	5.29	.09	3.527	6
25	24	19 13.2	89 51	6.63	.12	3.527	6

No. 2. Σ 2379, Dist. 13″. Observed as a single star. 6.00 is the mean of two measures (6.12, 6.06) observed with twenty extinctions 1883.519 and with six, 1885.437. H.P. = 5.75.
Nos. 3, 5. Spectrum I a! (Vogel.)
No. 6. Σ 2424. The faint distant companion not observed.
No. 8. Suspected variable by various authorities.
No. 11. Marked red by Gould : colour not salient.
No. 17. Decided variability stated by Gillis, but not confirmed.   See Ast. Obs. p. 669.)
No. 21. Σ 2489, Dist. 8″. Observed as one object. Spectrum I a ! (Vogel.)
No. 23. Spectrum I a ! (Vogel.)
No. 25. Σ 401. The distant companion (423″) not observed.

Reference Number.	Star's Designation.	R.A. 1890.	N.P.D. 1890.	Adopted Zenithal Magnitude. Polaris 2.05.	Average Deviation in Magnitude.	Date, 1880 +	Mag. Argel. Uran.
		h. m.	° ′				
26	28	19 14.5	77 50	5.55	0.04	3.565	6
27	26	19 14.7	95 37	5.21	.07	3.554	5
28	27	19 14.9	91 6	5.74	.02	3.554	6
29	L. 36489	19 16.7	90 29	5.86	.02	3.595	6
30	31	19 19.7	78 18	5.33	.05	3.565	5...6
31	δ.................	19 19.9	87 6	3.36	0.07	2.586	3...4
32	ν.................	19 20.9	89 53	5.08	.03	...	5
33	L. 36715	19 21.3	77 11	5.93	.05	3.604	6
34	L. 36781	19 22.5	75 57	6.08	.07	3.604	6
35	35	19 23.5	88 16	5.45	.06	3.565	6...5
36	L. 36867	19 24.5	75 38	5.69	0.03	3.604	6
37	36	19 24.9	93 1	5.23	.03	3.565	5...6
38	μ	19 28.7	82 51	5.10	.06	...	5...4
39	37	19 29.1	100 48	5.55	.05	3.595	6...5
40	κ	19 31.0	97 16	5.01	.04	3.595	5
41	ι	19 31.0	91 32	4.52	0.03	2.586	4...5
42	L. 37191	19 31.7	78 58	6.25	.03	3.604	5
43	42	19 32.0	94 54	5.67	.11	3.565	6
44	σ	19 33.8	84 51	4.88	.04	3.595	5
45	45	19 35.1	90 53	5.48	.06	3.565	6
46	W.B. 19ʰ-884 ...	19 36.0	76 26	6.09	0.04	3.609	6
47	χ	19 37.4	78 26	5.60	.06	3.609	6
48	ψ	19 39.5	76 58	6.06	.05	3.609	6
49	υ.................	19 40.3	82 39	6.06	.04	...	6
50	γ	19 41.0	79 39	2.81	.05	2.745	3

Nos. 26, 28, 31. Spectrum I a! (Vogel.)

No. 29. Called red in Uran. Argent.

No. 32. 5.08 is the mean of three closely accordant observations, made 1883.595, and 1885.434 and .437. H.P. = 4.80.

No. 37. Birm., No. 500. The star is orange tinted.

No. 38. 5.10 is the mean of three closely accordant observations, made 1882.568, 1885.434 and .437. There is strong suspicion of variability. (See Uranometria Argentina, p. 323.)

No. 41. Various authorities give discordant magnitudes. (Variable?)

Nos. 45, 46. Spectrum I a! (Vogel.)

No. 49. 6.06 is the mean of three closely accordant magnitudes, made 1883.565, and 1885.434 and .437. Spectrum I a! (Vogel.)

No. 50. Birm., No. 512. Spectrum II a!! (Vogel.)

Reference Number.	Star's Designation.	R.A. 1890.	N.P.D. 1890.	Adopted Zenithal Magnitude. Polaris 2.05.	Average Deviation in Magnitude.	Date, 1880+	Mag. Argel. Uran.
		h. m.	° ′				
51	Bradley 2555 ...	19 43.0	101 9	6.30	0.07	3.571	6
52	π	19 43.5	78 27	5.62	.08	3.584	6
53	51	19 44.7	101 3	5.61	.06	3.571	6
54	α................	19 45.4	81 26	1.04	.08	...	1...2
55	L. 37763	19 45.5	92 44	6.42	.06	3.571	6
56	o................	19 45.7	79 52	5.13	0.02	3.571	6...5
57	η	19 46.9	89 17	4.22	.09	3.589	Var.
58	L. 37832	19 47.5	93 24	5.65	.02	3.571	5
59	57	19 48.7	98 31	5.42	.05	3.571	5
60	ξ................	19 48.9	81 49	4.86	.08	3.574	5
61	58	19 49.1	90 1	5.74	0·07	3.574	6
62	β	19 49.9	83 52	3.69	.06	...	4
63	φ	19 51.0	78 52	5.60	.03	3.571	5...6
64	L. 38100	19 53.8	100 14	6.03	.03	3.574	6
65	L. 38199	19 55.7	81 44	6.22	.05	3.582	6
66	62	19 58.7	91 1	5.91	0.03	3.582	6
67	τ................	19 58.8	83 2	5.75	.03	3.585	6...5
68	L. 38506	20 2.6	80 56	6.40	.02	3.582	6
69	L. 38544	20 3.4	79 36	6.21	.03	3.582	6
70	θ	20 5.6	91 9	3.26	.03	2.589	3
71	W.B. (2) 20ʰ−105	20 7.0	89 29	6.47	0.03	3.595	6
72	66	20 7.6	91 20	5.77	.07	3.584	6
73	ρ................	20 9.2	75 8	5.07	.07	3.609	5
74	W.B. (2) 20ʰ−295	20 14.0	91 26	6.26	.08	3.601	6
75	L. 39102	20 15.3	72 34	5.95	.02	3.609	6...5
76	B.A.C. 7014	20 17.7	85 1	5.32	0.03	3.601	6 ..5
77	L. 39224	20 19.0	89 17	6.14	.06	3.609	6
78	68	20 22.7	93 43	6.10	.08	3.584	6
79	69	20 23.9	93 15	5.16	.08	3.584	5
80	L. 39540	20 26.7	88 9	6.54	.08	3.601	6

No. 52. Σ 2583, Dist. 1″.5. Consequently observed as one star.

No. 54. 1.04 is the mean of five accordant determinations made on as many nights, two at Oxford and three at Cairo, involving sixty extinctions made between 1882.750 and 1883.170. H.P. = 0.97. Spectrum I a !! (Vogel.)

No. 57. Varies from 3.5 to 4.7 mag. Period 7ᵈ, 4ʰ 14ᵐ.4ˢ, but some of the recently observed minima, according to Schönfeld, are not well represented.

No. 59. Σ 2594, Dist. 32″. The brighter star observed.

No. 62. 3.69 is the mean of three closely accordant observations, made 1882.589, and 1885.434 and .437. There is some ground for suspecting variability.

Nos. 63, 70, 73. Spectrum I a ! (Vogel.)

No. 68. Σ 2628, Dist. 4″. The observation is of the combination.

Reference Number.	Star's Designation.	R.A. 1890.	N.P.D. 1890.	Adopted Zenithal Magnitude. Polaris 2.05.	Average Deviation in Magnitude.	Date, 1880+	Mag. Argel. Uran.
		h. m.	o '				
81	L. 39542	20 26.8	88 15	6.32	0.01	3.609	6
82	70	20 31.0	92 56	5.28	.06	3.584	5
83	71	20 32.7	91 29	4.63	.10	2.589	4...5

No. 82. Birm., No. 564.

ARIES.

1	4	1 42.2	73 35	5.63	0.04	4.078	6
2	1...................	1 44.1	68 16	6.20	.05	4.007	6
3	P. i. 185..........	1 45.0	79 30	5.87	.05	4.007	6
4	γ ....:........	1 47.5	71 15	4.11	.04	2.679	4...3
5	β	1 48.6	69 44	2.75	.05	2.679	3...2
6	ι....................	1 51.3	72 43	5.21	0.04	4 007	6
7	λ	1 51.8	66 56	4.97	.02	4.007	5
8	P. i. 223..........	1 53.5	78 14	6.23	.04	4.075	6
9	10	1 57.4	64 36	5.82	.08	4.078	6
10	κ....................	2 0.4	67 52	5.31	.04	4.078	6...5
11	α	2 1.0	67 3	2.13	0.06	...	2
12	14	2 3.2	64 35	4.97	.06	4.089	5
13	15	2 4.5	71 1	5.64	.04	4.078	6
14	η..................	2 6.6	69 18	5.50	.03	4.007	5...6
15	19	2 7.1	75 14	5.98	.04	4.078	6
16	20	2 9.5	64 44	5.80	0.06	4.089	6
17	21	2 9.5	65 28	5.69	.03	4.089	6
18	θ	2 12.0	70 36	5.50	.04	4.078	6...5
19	ξ..................	2 18.9	79 53	5.48	.04	4.078	5...6
20	D.M.+22°, No. 354	2 23.0	67 1	6.11	.08	4.089	6

Nos. 1, 8, 18. Spectrum I a ! (Vogel.)
No. 2. Σ 174, Dist. 2".4. Observed as one mass.
No. 3. B.A.C. assigns this star to Cetus.
No. 4. Σ 180, Dist. 9". Observed as one mass. Spectrum I a ! (Vogel.)
No. 7. Σ unnum., Dist. 37". The brighter star observed.
No. 9. Σ 208, Dist. 1".4. Binary. Observed as one mass.
No. 11. Birm., No. 35. The colour is yellow. 2.13 is the mean of six accordant determinations, made on as many nights, three at Oxford and three at Cairo, involving seventy extinctions made between 1882.679 and 1883.175.
No. 13. Spectrum III a ! (Vogel.)
No. 16. Birm., No. 38. The red colour is not salient.

Refer- ence Number.	Star's Designation.	R.A. 1890.	N.P.D. 1890.	Adopted Zenithal Magnitude. Polar's 2.05.	Average Deviation in Magnitude.	Date, 1880 +	Mag. Argel. Uran.
		h. m.	° ′				
21	P. ii. 96	2 24.2	65 15	5.78	0.03	4.089	6
22	26	2 24.5	70 38	6.20	.04	4.075	6
23	30 1st star	2 30 6	65 50	6.33	.04	4.089	} 6
24	30 2nd star	2 30.7	65 50	6.78	.10	4.089	
25	31	2 30.7	78 2	5.68	.04	4 078	6
26	ν...............	2 32 6	68 31	5.54	0 09	4.078	6...5
27	33	2 34 3	63 25	5.58	.07	4.089	6 . 5
28	μ	2 36.2	70 27	5.05	.05	4.089	6 . 5
29	35	2 37.0	62 46	4.51	.04	4.089	5
30	ο	2 38 5	75 9	5.08	.05	4 089	6
31	38	2 39.0	78 1	5.40	0.06	4.114	5
32	39	2 41 4	61 12	4.65	.05	5.124	5
33	P. ii. 181	2 42.4	65 16	6.09	.11	4.114	6 5
34	π	2 43.2	73 0	5.51	.10	4.114	6...5
35	41	2 43 5	63 11	3.58	.08	2.704	4
36	σ	2 45 4	75 22	5.58	0.10	4.114	6
37	45	2 49 6	72 7	5.72	.06	4.114	...
38	ρ...............	2 50.2	72 25	5.08	.08	4 114	6
39	ε...............	2 52 9	69 6	4.24	.07	2.704	4...5
40	49	2 55 4	63 58	6.09	.08	4.114	6
41	52	2 59 0	65 10	5.26	0 05	4.114	6
42	55	3 3.0	61 21	5.58	.02	4.114	6
43	δ...............	3 5.3	70 41	4.54	.04	2.704	4...5
44	56	3 5.7	63 9	5.52	.10	4.114	6
45	ζ	3 8.6	69 22	4.78	.08	2.704	4...5

Nos. 22 and 28. Spectrum I a! (Vogel.)
Nos. 23 and 24. These stars form Σ 5¹, Dist. 38″.
No. 27. Σ 289, Dist. 28″. The bright star observed.
No. 31. = 88 Ceti. (See Introd. to B.A.C., p. 75.) Spectrum I a! (Vogel.)
No. 34. Σ 311. Triple. The closer pair, distance 3″.1, observed as one mass. Considered by Struve to be variable. (See Mens. Microm., p. lxxii.)
No. 35. Σ unnum. The brighter star observed.
No. 37. Spectrum III a!!! (Vogel.)
No. 39. Σ 333, Dist. 1″.3. Observed as one mass. Probably variable. (See Mens. Microm., p. lxxii. and Ast. Nach. LXX., p. 317.) An observation on 1885.635 gave 4.51 as the magnitude for the combined light of the stars. H.P. = 4.58.
No. 41. Σ 346. Quadruple. The closer pair, distance 1″, observed as one mass.

Reference Number.	Star's Designation.	R.A. 1890.	N.P.D. 1890.	Adopted Zenithal Magnitude. Polaris 2.05.	Average Deviation in Magnitude.	Date, 1880 +	Mag. Argel. Uran.
		h. m.	° ′				
46	59	3 13.4	63 19	5.86	0.05	4.116	6
47	P. iii. 32	3 13.7	61 21	4.84	.09	4.116	5
48	τ¹	3 14.9	69 15	5.42	.06	...	5
49	62	3 15.6	62 47	5.60	.01	4.116	6
50	τ²	3 16.4	69 39	5.37	.04	4.116	5...6
51	64	3 17.8	65 40	5.66	0.09	4.116	6
52	65	3 18.1	69 35	5.45	.06	4.116	6

No. 48. **5.42** is the mean of two measures (5.44, 5.40) observed 1884.115 with twenty extinctions and 1885.635 with six. **H.P. = 5.18.**

AURIGA.

1	P. iv. 185	4 42.2	57 36	6.02	0.01	5.084	6
2	W.B. iv. 889	4 42.2	58 46	5.81	.09	5.133	6
3	1	4 42.5	52 42	5.27	.01	5.084	6
4	P. iv. 184	4 42.9	41 27	5.90	.05	5.084	6
5	R. 1328	4 45.0	47 36	5.85	.05		6
6	2	4 45.3	53 29	4.99	0.03	5.100	5
7	R. 1339	4 47.0	46 7	6.00	.09	5.133	6
8	ι	4 49.8	57 0	2.87	.08	2.865	3
9	4	4 51.8	52 17	5.32	.07	5.152	6
10	ε	4 54.1	46 20	3.04	.02	2.865	3...4
11	ζ	4 54.8	49 5	3.80	0.03	2.865	4
12	9	4 58.0	38 33	5.07	.05	5.084	5
13	η	4 58.8	48 55	3.49	.08	...	4...3
14	μ	5 5.9	51 39	5.15	.01	5.100	6 ..5
15	14	5 8.2	57 26	5.26	.06	5.133	5 ..6

No. 5. **5.85** is the mean of two measures (5.80, 5.90) observed 1885.992, and 1885.239, each with six extinctions. **H.P. = 5.54.**

No. 9. Σ 616, Dist. 7″. Observed as one mass.

No. 10. Variability discovered by Fritsch. Max. 3.0, Min. 4.5 mag. Period irregular.

No. 11. Birm., No. 93. The colour is slightly orange.

No. 13. **3.49** is the mean of six accordant determinations, made on as many nights, three at Oxford and three at Cairo, involving seventy extinctions made between 1882.865 and 1883.233. **H.P. = 3.33.**

No. 15. Σ 653, Dist. 14″. Observed as one mass. There is also a fainter companion.

Reference Number.	Star's Designation.	R.A. 1890.	N.P.D. 1890.	Adopted Zenithal Magnitude. Polaris 2.05.	Average Deviation in Magnitude.	Date, 1880+	Mag. Argel. Uran.
		h. m.	° '				
16	α	5 8.5	44 7	0.08	0.05	...	1
17	W.B. v. 162	5 9.0	55 49	6.04	.14	5 084	6
18	16	5 11.0	56 44	4.97	.12	5.138	5
19	λ	5 11.4	49 59	5.03	.11	5.100	5
20	P. v. 26	5 11.7	56 22	5.10	.03	5.089	6
21	W.B. v. 266	5 12.5	49 0	5.29	0.09	5.239	6
22	19	5 12.8	56 9	5.42	.07	...	5...6
23	ρ	5 14.0	48 18	5.38	.00	5.089	6...5
24	R. 1458	5 15.1	49 5	5.55	.10	5.152	6
25	σ	5 17.2	52 43	5.37	.03	5.084	6
26	P. v. 62	5 17.5	58 53	6.20	0.04	5.100	} 6
27	P. v. 63	5 17.5	58 58	5.83	.04	5.100	
28	L. 10143	5 20.1	59 54	5.84	.03	5.133	6
29	φ	5 20.4	55 36	5.44	.01	5 084	5...6
30	χ	5 25.5	57 53	5.08	.04	5.089	5
31	26	5 31.7	59 34	5.63	0.04	5.152	6
32	ο	5 37.4	40 13	5.70	.02	5.084	6...5
33	τ	5 41.5	50 51	4.83	.08	2.865	5
34	υ	5 43.5	52 43	5.13	.02	5.133	5
35	ν	5 43.9	50 53	4.20	.10	2.865	4
36	ξ	5 45.6	34 19	5.02	0.04	5.089	5
37	δ	5 50.5	35 43	3.98	.06	2.865	4...5
38	β	5 51.5	45 4	1.94	.05	2.865	2
39	π	5 51.8	44 4	4.68	.02	5.239	5
40	θ	5 52.2	52 48	3.03	.09	...	3

No. 16.   0.08 is the mean of seven accordant determinations, made on as many nights, four at Oxford and three at Cairo, involving eighty extinctions made between 1882.865 and 1883.205. H.P. = 0.18. Herschel and Σ think this star has increased in brilliancy.

No. 18.   OΣ 103, Dist. 4".5. Observed as one mass.

No. 19.   Σ 3^II, Dist. 104". The brighter star observed.

No. 21.   Birm., Add. I, No. 18. The star is slightly red.

No. 22.   5.42 is the mean of two measures (5.40, 5.44) observed 1885.133 and 1885.239, each with six extinctions. H.P. = 5.10.

No. 31.   Σ 753, Dist. 12". Observed as one mass. Variable? (See Gillis's Ast. Obs., p. 663.)

No. 35.   Birm., Add. I, No. 24. The red colour is not salient.

No. 36.   ξ Aurigæ = 32 Camelopardali. (See Introd. to B.A.C., p. 75.)

No. 37.   Birm., No. 129. The red colour is not salient.

No. 38.   Birm., The colour is orange.

No. 40.   3.03 is the mean of three accordant measures made 1882.865, 1885.636, 1885.639, involving thirty-two extinctions. H.P. = 2.67.

Refer-ence Number.	Star's Designation.	R.A. 1890.	N.P.D. 1890.	Adopted Zenithal Magnitude. Polaris 2.05.	Average Deviation in Magnitude.	Date, 1880+	Mag. Argel. Uran.
		h. m.	o '				
41	36	5 52.6	42 6	5.70	0.03	5.133	6
42	P. v. 280	5 54.2	40 6	6.18	.01	5.133	6
43	38	5 55.4	47 5	6.14	.01	5.089	6
44	40	5 59.0	51 30	5.61	.06	...	6
45	41	6 3.2	41 16	5.81	.03	5.089	6
46	κ	6 8.4	60 28	4.81	0.05	2.865	5...4
47	42	6 9.4	43 32	6.42	.02	5.152	} 6
48	43	6 10.1	43 36	6.44	.07	5.152	
49	45	6 12.8	36 30	5.83	.10	...	6
50	W.B. vi. 316	6 14.2	60 25	6.06	.16	5.182	6
51	ψ^1 46	6 16.4	40 39	5.22	0.04	...	5
52	48	6 21.5	59 26	5.48	.00	5.133	6...5
53	P. vi. 126	6 25.3	57 28	5.77	.02	5.149	6
54	49	6 28.3	61 53	4.95	.06	5.152	6...5
55	ψ^3 52	6 31.1	50 0	5.35	.01	5.149	6
56	53	6 31.4	60 56	5.43	0.07	5.638	6
57	ψ^2 50	6 31.5	47 25	5.09	.05	5.089	5
58	54	6 32.6	61 38	5.75	.09	5.182	6
59	ψ^4 55	6 35.1	45 22	5.23	.09	5.149	5
60	ψ^5 56	6 38.8	46 19	5.58	.11	5.182	6
61	ψ^6 57	6 39.3	41 6	5.32	0.01	5.089	6
62	ψ^7 58	6 43.0	48 5	5.11	.03	5.182	5
63	ψ^8 { 60	6 45.7	51 25	6.25	.07	5.149	} 6
64	61	6 46.4	51 21	6.05	.02	5.149	
65	ψ^9	6 48.4	43 34	5.96	.10	5.191	6

No. 44. 5.61 is the mean of three accordant measures made 1885.139, 1885.636, and 1885.639.
   H.P. = 5.30.
No. 45. Σ 845, Dist. 8″. Observed as one mass.
No. 46. Variable ? (See Gillis's Ast. Obs., p. 663.) H.P. = 4.46.
No. 49. 5.83 is the mean of three accordant measures made 1885.133, 1885.636, and 1885.639.
Nos. 51, 55, &c. The B.A.C. does not adopt the symbols ψ^1, ψ^2 ... ψ^{10}. The notation here given is that of Argelander.
No. 51. 5.22 is the mean of three accordant measures made 1885.089, 1885.636, and 1885.638. H.P. = 5.05.
No. 58. OΣ 152, Dist. 0.″9. Observed as one mass.
No. 60. Σ unnum., Dist. 37″. The larger star observed.

Refer-ence Number.	Star's Designation.	R.A. 1890.	N.P.D. 1890.	Adopted Zenithal Magnitude. Polaris 2.05.	Average Deviation in Magnitude.	Date. 1880 +	Mag. Argel. Uran.
		h. m.	° ′				
66	P. vi. 263 ψ^{10}......	6 49.6	44 46	4.98	0.09	5.191	5
67	P. vi. 316	6 58.9	55 21	5.73	.03	5.133	5...6
68	L. 13704	7 0.2	55 50	6.00	.08	5.191	6
69	63	7 4.1	50 30	5.05	.05	5.636	6
70	64	7 10.4	48 55	5.68	.05	5.191	6
71	65	7 14.7	53 2	5.55	0.07	5.191	6
72	66	7 16.5	49 7	5.43	.10	5.149	6

No. 67. The B.A.C. assigns this star to Gemini.

BOÖTES.

1	B.A.C. 4559	13 34.2	78 42	5.50	0.05	4.464	5
2	1	13 35.4	69 29	5.89	.05	4 450	6
3	2	13 35.8	66 57	5.87	.10	4.472	6
4	3	13 41.6	63 45	6.03	.04	4.472	6
5	τ	13 42.1	72 0	4.58	.08	5.327	5...4
6	v	13 44.2	73 39	3.91	0.05	2.370	4...5
7	6 (*e*)	13 44.5	68 11	4.93	.06	4.483	5
8	P. xiii. 225	13 46.9	77 17	6.02	.02	4.472	6
9	η	13 49.5	71 3	2.74	.05	...	3
10	9	13 51.6	61 58	4.97	.03	4.450	5
11	P. xiii. 264	13 53.4	74 49	5.65	0.05	4.472	6
12	10	13 53.5	67 46	5.24	.11	4.472	6
13	L. 25746	13 55.9	80 34	6.00	.11	4.464	6
14	11	13 56.2	62 5	6.06	.02	4.483	6
15	P. xiii. 316	14 2.5	45 37	5.47	.03	4.475	6...5

No. 1. The B.A.C. assigns this star to Virgo.

No. 2. This star is Σ 1772, Dist. 5″. The observation refers to the brighter component.

No. 5. The observation of 1882.370 (see Mem. R.A.S., vol. xlvii., p. 423) is rejected; the sky was hazy and varying meteorologically.

No. 6. Noted by Schmidt and Argelander as red. Birmingham as pale yellow. The red colour was not noticeable at the time of observation.

Nos. 8, 9. and 13. Spectrum I a ! (Vogel.)

No. 9. 2.74 is the mean of three accordant determinations made 1882.370, 1883.173, and 1883.183.

Reference Number.	Star's Designation.	R.A. 1890.		N.P.D. 1890.		Adopted Zenithal Magnitude. Polaris 2.05.	Average Deviation in Magnitude.	Date, 1880+	Mag. Argel. Uran.
		h.	m.	°	′				
16	13	14	4.2	40	1	5.68	0.06	4.445	6
17	12 (d)	14	5.4	64	23	4.81	.04	4.450	5
18	14	14	8.8	76	31	5.66	.05	...	6
19	15	14	9.5	79	23	5.45	.11	4.464	6
20	κ	14	9 5	37	42	4.56	.09	2.444	4...5
21	B.A.C. 4728	14	10.0	47	58	6.21	0.06	4.445	6
22	α	14	10.7	70	13	0.31	.05	...	1
23	B.A.C. 4738	14	11.9	49	45	6.18	.06	4.445	6
24	λ	14	12.2	43	25	4.16	.04	2.444	4
25	ι	14	12.3	38	8	4.63	.05	2.444	4...5
26	P. xiv. 45 (A) ...	14	13.4	53	59	4.98	0.05	4.445	5
27	18	14	13.9	76	29	5.24	.05	4.450	6
28	20	14	14.6	73	11	4.65	.09	4.464	5
29	B.A.C. 4758	14	15.3	50	42	6.19	.08	4.445	6
30	D.M.+26°, No. 2554	14	15.8	63	26	6.91	.11	4.445	6
31	P. xiv. 69	14	18.0	81	3	5.29	0.06	4.464	5...4
32	P. xiv. 73	14	18.7	83	41	5.25	.01	4.464	5
33	22 (f)	14	21.4	70	17	5.15	.06	4.483	5
34	θ	14	21.5	37	38	4.02	.06	2.444	4...3
35	24 (g)	14	24.8	39	41	5.67	.03	4.445	6
36	ρ	14	27.1	59	9	3.56	0.09	2.444	4...3
37	B.A.C. 4809	14	27.5	62	51	6.05	.03	4.450	6
38	26	14	27.6	67	15	5.91	.06	4.483	6
39	γ	14	27.7	51	13	3.21	.08	...	3...2
40	σ	14	29.9	59	47	4.62	.03	4.483	5...4

No. 16. This star is Birmingham No. 320. The colour is decidedly orange.

No. 18. This magnitude (5.66) is the mean of two measures 5.72 and 5.60 observed with twenty extinctions 1884.464, and with six in 1885.327. **H.P. = 5.46.**

No. 20. This star is Σ 1821, Dist. 13″. The brighter component is observed.

No. 22. **0.31** is the mean of seven accordant determinations, made on as many nights, four at Oxford and three at Cairo, involving eighty extinctions made between 1882.370 and 1883.200. **H.P. = 0.03.**

No. 25. This star is Σ 3124, Dist. 38″. The fainter component is not observed.

No. 30. Several small stars of approximately equal magnitude near. The star in the text accords best with Argelander's place.

No. 31. This star is Σ 1835, Dist. 6″. The brighter component observed. Spectrum I a! (Vogel.)

No. 39. **3.21** is the mean of four accordant determinations, made 1882.444, 1883.183, 1883.186, and 1883.192.

Reference Number.	Star's Designation.	R.A. 1890.	N.P.D. 1890.	Adopted Zenithal Magnitude. (Polaris 2.05.	Average Deviation in Magnitude.	Date. 1880+	Mag. Argel. Uran.
		h. m.	o ´				
41	B.A.C. 4830	14 30.8	40 9	6.10	0.02	...	6
42	33	14 34.8	45 7	5.24	.08	4.450	6
43	P. xiv. 156.........	14 34.8	35 30	5.75	.04	4.475	6
44	π	14 35.6	73 7	4.64	.11	5.327	4
45	ζ..................	14 35.9	75 48	3.88	.03	5.327	3...4
46	31	14 36.2	81 22	5.04	0.10	4.464	5...4
47	32	14 36.5	77 52	5.74	.06	4.472	6
48	34	14 38.6	63 0	4.76	.01	4.483	6
49	o	14 40.1	72 34	4.56	.10	2.370	5...4
50	P. xiv. 178.........	14 40.9	74 24	6.07	.05	4.472	6
51	ε¹	14 40.2	62 28	2.47	0.05	2.449	} 2...3
52	ε²	14 40.2	62 28	5.03	.08	2.449	
53	W.B. xiv. 908	14 43.5	65 11	6.27	.04	4.483	6
54	L. 27017	14 44.3	81 33	6.95	.10	4.464	6
55	B.A.C. 4897	14 44.8	51 44	6.02	.12	4.475	6
56	P. xiv. 193.........	14 45.2	60 56	5.98	0.03	4.483	6
57	W.B. xiv. 945	14 45.3	65 37	6.00	.04	4.483	6
58	38 (h)	14 45.4	43 25	5.79	.03	4.475	6
59	39	14 46.0	40 50	5.51	.06	4.486	6
60	B.A.C. 4906	14 46.2	52 17	5.40	.08	4.486	6
61	ξ	14 46.3	70 27	4.58	0.08	2.444	4
62	B.A.C. 4926	14 51.0	75 7	5.60	.12	5.327	6
63	B.A.C. 4933	14 52.1	73 10	5.93	.05	4.464	6
64	B.A.C. 4937	14 52.7	39 55	5.79	.07	4.489	6
65	40	14 55.9	50 18	5.44	.09	4.489	5

No. 41. This magnitude (6.10) is the mean of two measures (6.12 and 6.08) observed with twenty extinctions 1884.475, and with six extinctions 1885.327. H.P. = 5.84.

No. 44. This star is Σ 1864, Dist. 6″. The observation of 1882.370 is rejected. (See Note to No. 5.)

No. 45. This star is Σ 1865, Dist. 1″. The magnitude is consequently that of the combination. The observation 1882.370 is rejected. (See Note to No. 5.)

No. 46. Gould suspects this star of variability. But the measure at Harvard, 4.99, accords with the Oxford measure here given.

No. 48. Discovered by Schmidt to be variable in 1867, with limits of magnitude 5.2–6.1. Schmidt's estimated maximum is too faint.

No. 50. Spectrum III a !!! (Vogel.)

No. 51. This star is Birmingham No. 339 and Σ 1877. The colours recorded vary from orange to yellow.

No. 53. This star is Σ 1884, Dist. 1″.2. Observed as one star.

No. 54. Gould's suspicion of the variability of this star is not supported by the Harvard and Oxford observations, but the magnitude 6.95 seems too faint for its detection by Argelander.

No. 61. Σ 1888, Binary. The larger component observed.

Refer-ence Number.	Star's Designation.	R.A. 1890.	N.P.D. 1890.	Adopted Zenithal Magnitude. Polaris 2.05.	Average Deviation in Magnitude.	Date, 1880 +	Mag. Argel. Uran.
		h. m.	° ′				
66	ω	14 57.3	64 33	4.70	0.09	2.449	5...4
67	β	14 57.8	49 10	3.64	.05	2.449	3
68	B.A.C. 4961	14 58.7	54 22	5.78	.03	4.494	6
69	ψ	14 59.7	62 37	4.45	.04	2.452	4...5
70	44	15 0.2	41 55	4.65	.10	2.452	5
71	47 (k)	15 1.8	41 25	5.70	0.06	4.494	5...4
72	W.B. xiv. 1327	15 2.3	53 7	6.19	.08	4.489	6
73	45 (c)	15 2.5	64 42	5.13	.09	4.483	5...4
74	46 (b)	15 3.7	63 17	5.05	.03	4.483	6
75	W.B. xv. 106	15 7.1	70 37	5.89	.04	4.464	6
76	P. xv. 18	15 8.7	66 35	6.17	0.07	4.486	6
77	χ	15 9.9	60 26	5.47	.05	4.486	5
78	δ	15 11.1	56 16	3.44	.13	2.452	3
79	P. xv. 36	15 13.5	69 1	5.52	.02	4.486	6
80	50	15 17.4	56 40	5.57	.02	4.489	5...6
81	R. 3369	15 18.6	50 1	5.81	0.04	4.489	6
82	μ	15 20.4	52 14	4.52	.08	2.452	4...3
83	P. xv. 81	15 22.0	55 17	6.50	.03	4.486	6
84	W.B. xv. 555	15 26.3	52 50	6.35	.05	4.486	6
85	ν¹	15 27.0	48 47	4.64	.10	...	4
86	W.B. xv. 581	15 27.2	53 1	6.30	0.02	4.486	6
87	ν²	15 27.8	48 44	4.98	.08	2.452	4
88	φ	15 33.9	49 17	5.24	.09	4.486	5

No. 70. Σ 1909, Binary. The magnitude is that of the combination.
No. 75. Spectrum III a !! Vogel.)
No. 82. This star is Σ 1938, Binary. The magnitude is that of the combined stars.
No. 85. This magnitude (4.64) is the mean of two measures (4.61 and 4.67) observed with twenty extinctions 1882.452, and with six extinctions 1885.327. H.P.=5.00.

CAMELOPARDALUS.

Reference Number.	Star's Designation.	R.A. 1890.	N.P.D. 1890.	Adopted Zenithal Magnitude. Polaris 2.05.	Average Deviation in Magnitude.	Date. 1880 +	Mag. Argel. Uran.
		h. m.	° ′				
1	B.A.C. 1001	3 10.3	24 45	4.68	0.01	5.237	6
2	P. iii. 51	3 20.2	30 26	4.55	.09	...	5...4
3	B.A.C. 1062	3 21.1	31 30	4.74	.09	5.234	5
4	B.A.C. 1065	3 21.6	34 56	5.07	.12	5.234	5
5	B.A.C. 1111	3 32.6	27 8	5.21	.04	5.256	6
6	B.A.C. 1127	3 35.6	23 8	5.71	0.10	5.236	6
7	P. iii. 105	3 36.4	27 0	5.20	.09	5.234	6...5
8	R. 1042	3 37.8	19 28	5.49	.15	5.252	5
9	γ	3 38.6	19 0	4.54	.05	2.812	4...5
10	P. iii. 121	3 39.4	24 49	4.75	.05	5.234	5
11	P. iii. 177	3 47.7	27 15	5.03	0.04	...	6...5
12	P. iii. 178	3 47.7	29 13	5.24	.04	5.234	6
13	P. iii. 208	3 55.3	31 9	5.14	.10	5.234	6...5
14	O.A. 4458	4 1.7	18 10	6.22	.11	5.237	6
15	P. iii. 260	4 7.2	28 26	5.63	.06	5.270	6
16	P. iv. 7	4 8.1	36 40	5.11	0.04	5.234	5
17	P. iv. 10	4 10.3	25 8	5.36	.08	5.239	6
18	P. iv. 22	4 12.2	29 31	5.72	.06	5.280	6...5
19	B.A.C. 1318	4 12.9	33 45	6.12	.07	5.234	6
20	L. 7984	4 13.6	30 38	6.00	.01	5.256	6
21	R. 1221	4 20.7	17 42	5.91	0.05	5.270	6
22	1	4 23.3	36 20	5.48	.16	5.234	6
23	O.A. 4895	4 26.1	25 59	5.97	.03	5.237	6
24	2	4 31.2	36 45	5.48	.05	5.301	6
25	3	4 31.2	37 8	5.78	.06	...	6

No. 1. B.A.C. assigns this star Cassiopeia.

No. 2. Σ 385, Dist. 2″. Observed as one mass. 4.55 is the mean of two measures (4.63, 4.47) observed with twenty extinctions 1885.234, and with six 1885.382. H.P. = 4.10.

No. 4. Σ 390, Dist. 15″. The faint companion not observed.

No. 10. Birm., No. 69. The colour is not salient.

No. 11. 5.03 is the mean of two measures (5.05, 5.01) observed with twenty extinctions 1885.253, and with six 1885.301. H.P. = 4.75.

No. 12. Birm., No. 71. The red colour is not salient.

No. 22. Σ 550, Dist. 10″. Observed as one mass.

No. 24. Σ 566, Dist. 1″.5. Observed as one star.

No. 25. 5.78 is the mean of two measures (5.71 and 5.84) observed with twenty extinctions 1885.234, and with six on 1885.301. H.P. = 5.37.

Reference Number.	Star's Designation.	R.A. 1890.	N.P.D. 1890.	Adopted Zenithal Magnitude. Polaris 2.05.	Average Deviation in Magnitude.	Date, 1880 +	Mag. Argel. Uran.
		h. m.	o '				
26	P. iv. 112	4 34.0	14 16	5.86	0.11	5.256	6
27	4	4 38.8	33 26	5.41	.10	5.239	6
28	Bradley 651	4 39.1	34 33	6.05	.10	5.280	6
29	P. iv. 170	4 41.8	26 38	5.76	.09	5.280	6
30	a	4 43.2	23 50	4.46	.07	2.812	4
31	5..................	4 46.1	34 55	5.82	0.09	5.234	6
32	7..................	4 48.5	36 25	4.72	.10	5.234	5
33	P. iv. 207	4 50.8	16 6	5.97	.02	5.280	6
34	β	4 53.8	29 43	4.27	.04	2.812	4
35	11	4 56.6	31 11	5.19	.02	5.234	5
36	12 ,................	4 56.6	31 8	5.74	0.04	5.234	...
37	P. iv. 254	4 58.5	16 12	5.59	.06	5.270	5...6
38	P. iv. 269	5 4.4	10 54	5.15	.09	5.280	5
39	B.A.C. 1585	5 4.6	16 51	5.49	.10	5.270	6
40	15	5 10.0	32 0	6.21	.04	5.234	6
41	P. iv. 317	5 12.5	12 7	6.39	0.07	5.256	6
42	16	5 14.0	32 34	5.14	.02	5.234	6...5
43	17	5 19.8	27 1	5.58	.01	5.280	6
44	B.A.C. 1751	5 31.4	24 22	5.87	.07	5.237	6
45	B.A.C. 1813	5 41.1	21 34	6.22	.04	5.301	6
46	31	5 45.1	30 8	5.52	0.05	...	5...6
47	37	6 0.3	31 3	5.32	.05	5.301	6
48	36	6 1.8	24 16	5.31	.07	5.301	6...5
49	40	6 5.8	29 58	5.74	.07	5.239	6
50	P. v. 335	6 6.7	20 38	4.60	.07	2.708	5...4
51	R. 1707	6 15.7	19 24	6.04	0.06	5.261	6
52	R. 1723	6 21.4	10 19	6.09	.07	5.270	6
53	B.A.C. 2069	6 23.4	11 55	5.85	.09	5.261	6
54	P. vi. 75	6 27.5	10 18	5.71	.03	5.301	5...6
55	O.A. 6978	6 27.5	18 10	6.15	.06	5.237	6

No. 32. Σ 610, Dist. 26″. The larger star observed.
No. 38. Σ 634, Dist. 34″. The larger star observed.
No. 46. 5.52 is the mean of two measures (5.47, 5.58) observed with twenty extinctions 1885.234, and with six on 1885.301. H.P. = 5.23.

Reference Number.	Star's Designation.	R.A. 1890.	N.P.D. 1890.	Adopted Zenithal Magnitude. Polaris 2.05.	Average Deviation in Magnitude.	Date, 1880 +	Mag. Argel. Uran.
		h. m.	° ′				
56	42	6 39.5	22 18	4.92	0.04	5.261	5
57	43	6 41.8	21 0	5.04	.09	5.239	5
58	P. vi. 201	6 44.0	12 53	4.67	.06	2.708	5...4
59	R. 1854	6 53.4	19 6	6.27	.02	5.256	6
60	R. 1882	7 1.2	18 0	6.29	.10	5.237	6
61	P. vi. 292	7 8.0	7 22	5.20	0.04	5.253	5
62	B.A.C. 2419	7 16.5	23 26	5.94	.06	...	6
63	P. vii. 67	7 19.4	21 19	5.70	.02	5.253	6
64	51	7 36.1	24 16	5.00	.04	5.237	6
65	P. vii. 132	7 37.2	9 27	6.29	.08	5.239	6...5
66	B.A.C. 2596	7 47.0	10 13	5.30	0.02	5.239	5
67	P. vii. 187	7 47.0	15 47	5.51	.06	5.253	6...5
68	B.A.C. 2585	7 50 6	5 26	6.28	.09	5.280	6
69	R. 2057	7 52.1	30 39	6.26	.03	5.280	6
70	53	7 52.3	29 25	6.01	.07	5.234	6
71	D.M.+63°, No. 749	7 52.8	26 37	6.02	0.04	5.234	6
72	54	7 53.6	32 25	6.37	.08	5.239	6
73	R. 2092	8 5.7	13 54	5.64	.06	5.253	5
74	B.A.C. 2852	8 27.4	15 59	6.29	.08	5.239	6
75	R. 2218	8 52.3	5 23	6.15	.08	5.280	6
76	R. 2312	9 16.3	14 26	6.26	0.05	5.253	6
77	P. ix. 252	10 13.6	5 11	5.34	.10	5.270	6
78	P. x. 22	10 17.6	6 53	5.08	.08	5.253	5
79	B.A.C. 3906	11 24.2	8 16	6.00	.01	5.280	6
80	B.A.C. 4106	12 6.3	7 41	6.12	.07	5.253	6
81	Brad. 1730........	12 48.2	5 59	5.71	0.09	2.708	} 5...4
82	Brad. 1731........	12 48.3	5 59	5.05	.08	2.708	
83	P. xiii. 133........	13 26.0	10 47	6.00	.09	5.253	6
84	P. xiii. 263........	13 45.5	6 42	6.10	.04	5.253	6
85	B.A.C. 4982	14 57.6	7 1	5.76	.01	5.253	6

No. 59. Birm., No. 160. The red colour is not salient.
No. 61. Birm., No. 168. Noted to be yellow rather than red.
No. 62. 5.94 is the mean of two measures (5.92, 5.96) observed with twenty extinctions 1885.256, and with six 1885.301.
No. 67. Birm., No. 192. The red colour is not salient.
Nos. 68, 80, 81, 82, 83, 84, and 85. B.A.C. assigns these stars to Ursa Minor.
No. 74. B.A.C. assigns this star to Ursa Major.
Nos. 81 and 82. These stars form Σ 1694, Dist. 22″.

CANCER.

Reference Number.	Star's Designation.	R.A. 1890.	N.P.D. 1890.	Adopted Zenithal Magnitude. Polaris. 2.05.	Average Deviation in Magnitude.	Date, 1880 +	Mag. Argel. Uran.
		h. m.	° '				
1	ω	7 54.3	64 18	6.03	0.05	4.305	6
2	8	7 59.0	76 34	5.06	.05	4.300	6
3	μ	8 1.5	68 6	5.55	.05	...	6...5
4	ψ	8 3.8	64 9	6.02	.04	...	6
5	ζ	8 5.9	72 1	4.94	.04	3.019	5...4
6	β	8 10.6	80 28	3.70	0.03	5.272	4...3
7	χ	8 13.4	62 25	5.26	.04	4.305	6
8	P. viii. 42	8 13.9	68 54	5.80	.06	5.272	6
9	λ	8 14.0	65 38	5.88	.06	5.272	6
10	20	8 17.1	71 19	6.09	.05	4.300	6
11	d^2	8 19.6	72 35	6.19	0.03	5.272	6
12	ϕ^1	8 19.8	61 44	6.09	.07	4.305	6
13	ϕ^2	8 20.1	62 42	5.70	.05	4.305	6
14	P. viii. 67	8 20.1	82 5	5.38	.03	...	6
15	27	8 20.7	76 59	5.87	.04	4.305	6
16	29	8 22.5	75 26	6.05	0.05	4.305	6
17	v^3 (30)	8 25.0	65 33	5.92	.06	4.308	6
18	θ	8 25.3	71 32	5.89	.05	4.305	6
19	η	8 26.4	69 11	5.53	.03	4.308	6
20	32	8 26.5	65 32	6.46	.03	4.308	6
21	L. 16823	8 28.4	81 10	5.83	0.05	4.300	6
22	36	8 31.1	79 58	6.12	.04	4.300	6
23	γ	8 36.9	68 8	4.89	.10	2.320	4...5
24	45	8 37.2	76 55	5.67	.03	4.300	6
25	δ	8 38.4	71 26	4.27	.03	2.320	4

Nos. 2 and 10. Spectrum I a! (Vogel.)

No. 3. 5.55 is the mean of two measures (5.52, 5.58) observed 1884.305 with twenty extinctions, and 1885.263 with six. H.P. = 5.25.

No. 4. 6.02 is the mean of two measures (6.06, 5.97) observed 1884.305 with twenty extinctions, and 1885.272 with six. H.P. = 5.78. There is great discordance in the older recorded estimates.

No. 5. Σ 1196. Triple and orbital. Observed as one mass.

No. 6. The observation of 1882.320 (see Mem. R.A.S., vol. xlvii.) is rejected. The sky was hazy. Spectrum II a! (Vogel.)

No. 13. Σ 1223, Dist. 4".5. Observed as one mass.

No. 14. 5.38 is the mean of two measures (5.39, 5.37) observed 1884.300 with twenty extinctions, and on 1885.259 with six. H.P. = 5.09.

No. 15. Spectrum III a!! (Vogel.)

Reference Number.	Star's Designation.	R.A. 1890.	N.P.D. 1890.	Adopted Zenithal Magnitude. Polaris 2.05.	Average Deviation in Magnitude.	Date, 1880 +	Mag. Argel. Uran.
		h. m.	° ′				
26	49	8 38.8	79 31	5.74	0.04	4.305	6
27	ι	8 40.1	60 50	4.24	.05	3.019	4
28	50	8 40.9	77 29	6.05	.02	4.305	6
29	σ¹	8 45.8	57 7	5.67	.03	4.308	6
30	ρ¹	8 45.9	61 20	5.92	.02	4.305	6
31	57	8 47.5	59 0	5.60	0.04	4.308	6
32	ρ²	8 49.1	61 39	5.35	.03	4.300	6
33	59	8 50.2	56 40	5.45	.04	4.308	6
34	ο	8 51.1	74 15	5.53	.05	...	6
35	63	8 51.4	74 0	5.68	.04	4.300	6
36	α	8 52.5	77 43	4.36	0.00	...	4
37	64	8 52.8	57 9	5.51	.06	4.308	5
38	67	8 55.3	61 40	6.07	.04	4.314	6
39	ν	8 56.3	65 7	5.42	.06	4.308	6
40	τ	9 1.4	59 54	5.57	.09	4.314	6
41	κ	9 1.8	78 53	5.17	0.01	4.300	5
42	75	9 2.3	62 55	6.20	.02	4.308	6
43	ξ	9 3.2	67 31	5.18	.04	5.259	5
44	W.B. ix. 3	9 4.0	58 35	6.50	.02	4.314	6
45	π	9 9.2	74 36	5.66	.01	4.300	6
46	83	9 12.9	71 50	6.65	0.04	4.300	6

No. 27. Σ 1268, Dist. 30″. The brighter star only observed.
Nos. 34, 35, and 36. Spectrum I a! (Vogel.)
No. 34. 5.53 is the mean of two measures (5.56, 5.50) observed 1884.300 with twenty
extinctions, and 1885.259 with six. H.P. = 5.21.
No. 36. 4.36 is the mean of two identical measures observed 1882.320 with twenty extinctions,
and 1883.098 with ten.
No. 44. Birm., No. 218. The colour is noticeable. Strong suspicion of variability.

CANES.

Reference Number.	Star's Designation.	R.A. 1890.	N.P.D. 1890.	Adopted Zenithal Magnitude. Polaris 2.05.	Average Deviation in Magnitude.	Date, 1880+	Mag. Argel. Uran.
		h. m.	° ′				
1	2...................	12 10.6	48 44	5.96	0.06	4.355	6
2	P. xii. 29	12 11.0	56 19	5.15	.04	4.386	5
3	3...................	12 14.4	40 24	5.82	.03	4.355	6
4	4...................	12 18.4	46 51	6.05	.01	4.355	6
5	5...................	12 18.7	37 50	5.18	.04	4.386	5
6	6...................	12 20.4	50 22	5.23	0.03	4.344	5...6
7	R. 2866	12 22.1	48 2	6.82	.00	4.401	6
8	7...................	12 24.9	37 51	6.51	.04	4.401	6
9	R. 2876	12 25.6	36 19	6.40	.01	4.401	6
10	P. xii. 122........	12 28.2	56 8	5.54	.07	4.401	5
11	β	12 28.6	48 3	4.53	0.07	...	4...5
12	9...................	12 33.5	48 31	6.15	.02	4.355	6
13	W.B. xii. 683 ...	12 33.9	53 25	6.33	.04	4.404	6
14	B.A.C. 4282	12 39.3	45 18	6.27	.05	4.404	6
15	10	12 39.8	50 8	5.77	.01	4.404	6
16	B.A.C. 4287	12 40.0	43 58	5.27	0.01	4.442	5...6
17	R. 2914	12 42.8	39 15	6.68	.03	4.445	6
18	11	12 43.7	40 56	5.96	.03	4.344	6
19	B.A.C. 4311	12 45.0	51 52	6.06	.04	4.401	6
20	L. 24054	12 48.8	55 52	6.39	.05	4.404	6
21	B.A.C. 4341	12 49.9	42 12	5.84	0.06	4.355	6
22	α ,...............	12 50.9	51 5	3.32	.05	...	3
23	12 1st star	12 50.9	51 6	5.74	.07	2.449	
24	B.A.C. 4350	12 52.1	43 14	5.99	.04	4.442	6
25	14	13 0.6	53 37	5.42	.02	4.404	5

No. 1. Σ 1622, Dist. 11″. The faint star not observed.

No. 11. 4.53 is the mean of two measures (4.56 and 4.50) observed with twenty extinctions 1882.449, and with six 1885.382. H.P. = 4.30.

No. 16. Birm., No. 290. The star is decidedly red. Schmidt noted variability, and determined a period of about 386 days. A re-observation on 1885.382 gave 5.54 for the magnitude. H.P. = 5.59.

Nos. 22 and 23. These stars form Σ 1692, Dist. 20″. 3.32 is the mean of six determinations, made on as many nights, three at Oxford and three at Cairo, involving seventy extinctions made between 1882.449 and 1883.107. H.P. (the combination) = 3.00.

Reference Number.	Star's Designation.	R.A. 1890.		N.P.D. 1890.		Adopted Zenithal Magnitude. Polaris 2.05.	Average Deviation in Magnitude.	Date, 1880 +	Mag. Argel. Uran.
		h.	m.	°	′				
26	B.A.C. 4389	13	0.9	44	9	5.78	0.04	4.442	6
27	15	13	4.6	50	53	6.19	.06	4.445	
28	17	13	5.0	50	55	6.23	.04	4.445	} 5
29	P. xiii. 27	13	8.7	49	16	5.04	.05	4.355	5
30	19	13	10.6	48	34	5.91	.04	4.355	6
31	20	13	12.6	48	51	4.71	0.06	2.449	5...4
32	21	13	13.6	39	43	4.97	.02	4.401	5
33	23	13	15.4	49	16	5.83	.07	4.355	6...5
34	P. xiii. 71	13	16.0	45	26	6.13	.01	...	6
35	R. 3013	13	21.5	43	24	5.79	.06	5.382	6
36	R. 3017	13	23.6	48	42	6.26	0.08	4.404	6
37	B.A.C. 4519	13	26.5	47	20	6.13	.07	4.442	6
38	W.B. xiii. 557 ...	13	29.5	50	38	6.20	.02	4.442	6
39	P. xiii. 136.........	13	29.9	52	15	4.99	.05	4.404	5
40	24	13	30.0	40	25	4.97	.08	4.449	5
41	B.A.C. 4545	13	30.6	45	15	6.49	0.01	4.442	6
42	25	13	32.6	53	9	5.08	.02	5.382	5
43	W.B. xiii. 686 ...	13	35.3	58	26	6.22	.09	4.442	6
44	P. xiii. 163	13	35.5	61	22	6.47	.04	4.405	6
45	R. 3074	13	38.0	47	45	5.96	.03	...	6
46	R. 3079	13	38.9	43	56	6.69	0.05	4.445	6
47	B.A.C. 4596	13	41.6	48	22	5.61	.01	4.404	6...5
48	B.A.C. 4600	13	42.3	50	54	5.37	.06	4.448	5...6
49	B.A.C. 4609	13	43.4	47	29	6.40	.01	...	6
50	B.A.C. 4610	13	43.7	58	16	5.86	.06	4.448	6
51	B.A.C. 4627	13	46.2	54	41	5.99	0.04	4.448	6
52	B.A.C. 4628	13	46.3	54	47	6.35	.10	4.448	5
53	B.A.C. 4632	13	46.9	55	0	5.19	.09	4.448	...
54	P. xiii. 235	13	48.2	60	49	6.10	.03	4.448	6
55	B.A.C. 4652	13	50.8	57	26	6.64	.11	4.404	6
56	L. 25839	13	58.0	43	43	6.40	0.06	4.445	6

No. 34. 6.13 is the mean of two measures (6.13, 6.14) observed 1884.401 with twenty extinctions, and on 1885.382 with six. H.P. = 6.14. The star observed is R. 3002.

No. 42. Σ 1768. Close binary. Observed as one star.

No. 45. 5.96 is the mean of two measures (5.92, 6.01) observed on 1884.445 with twenty extinctions, and with six on 1885.382. H.P. = 6.44.

No. 49. 6.40 is the mean of two measures (6.43, 6.36) observed on 1884.445 with twenty extinctions, and 1885.382 with six. H.P. = 6.70.

No. 53. This star is not in Argelander, but it was probably the object described as No. 52.

CANIS MAJOR.

Reference Number.	Star's Designation.	R.A. 1890.	N.P.D. 1890.	Adopted Zenithal Magnitude. Polaris 2.05.	Average Deviation in Magnitude.	Date, 1880 +	Mag. Argel. Uran.
		h. m.	° '				
1	α	6 40.3	106 34	+1.95*	0.07	...	1

No. 1. **+1.95** is derived from five nights' observations at Cairo alone, between 1883.078 and 1883.111, involving fifty extinctions. The mean date is 1883.095. Further, observations on five nights at Oxford, between 1882.045 and 1883.117, when corrected for mean atmospheric absorption (Memoirs R.A.S., vol. xlvii, p. 414) gave +1.97. **H.P. = +2.43.**

* The symbol **+1.95** indicates 1.95 mag. brighter than a star of the first magnitude. (See Preface.)

CANIS MINOR.

1	P. vii. 8	7 6.0	84 10	6.15	0.04	4.209	6
2	1	7 18.9	78 7	5.56	.12	4.229	6
3	ε	7 19.6	80 30	5.09	.06	4.218	5...6
4	β	7 21.2	81 29	3.11	.05	...	3
5	η	7 22.1	82 50	5.66	.08	...	6
6	γ	7 22.2	80 51	4.77	0.11	4.209	5
7	6	7 23.7	77 46	4.87	.03	4.229	5
8	δ¹	7 26.4	87 51	5.03	.01	4.220	6
9	δ²	7 27.4	86 29	5.72	.11	4.220	6
10	α	7 33.5	84 29	+0.50*	.04	...	1
11	11	7 40.2	78 58	5.39	0.06	4.229	5...6
12	ζ	7 46.0	87 57	5.05	.04	4.220	6
13	14	7 52.7	87 29	5.26	.12	4.209	6
14	L. 15657	7 55.5	84 50	5.75	.05	4.220	6
15	P. vii. 289	7 56.6	87 22	4.41	.07	4.209	6

Nos. 1, 2, 4, 8, 10, 11, and 12. Each has the spectrum I a! (Vogel.)

No. 4. **3.11** is the mean of two measures (3.12, 3.09) observed 1882.188 with twenty extinctions, and 1883.215 with six.

No. 5. **5.66** is the mean of two measures (5.69, 5.63) observed 1884.218 with twenty extinctions, and on 1885.266 with six. **H.P. = 5.31.**

Nos. 6 and 7. Spectrum II a! (Vogel.)

No. 10. **+0.50** is the mean of seven accordant determinations made on as many nights, four at Oxford and three at Cairo, involving eighty extinctions, between 1882.188 and 1883.131. **H.P. = +0.54.**

* The explanation of this notation is given in the Preface.

No. 13. This star has two distant companions, not observed.

CASSIOPEIA.

Reference Number.	Star's Designation.	R.A. 1890.	N.P.D. 1890.	Adopted Zenithal Magnitude. Polaris 2.05.	Average Deviation in Magnitude	Date, 1880 +	Mag. Argel. Uran.
		h.　m.	°　′				
1	D.M.+56°. No. 2923	22　55.5	33　39	5.71	0.06	5.136	} 6
2	B.A.C. 8024	22　56.9	33　29	6.23	.04	5.136	
3	1..................	23　2.0	31　10	5.22	.01	...	5...6
4	2..................	23　5.0	31　16	5.57	.06	5.138	6
5	B.A.C. 8083	23　7.8	33　27	5.54	.01	5.136	6
6	4..................	23　20.0	28　19	5.49	0.06	...	6
7	P. xxiii. 101	23　24.9	32　3	4.94	.03	5.138	5
8	τ..................	23　41.7	31　58	5.09	.03	5.136	5
9	6..................	23　43.5	28　24	5.57	.02	5.138	6
10	ρ..................	23　48.9	33　7	4.80	.07	2.835	5
11	P. xxiii. 237	23　51.6	34　54	5.55	0.03	5.138	6
12	σ	23　53.4	34　52	5.02	.05	5.109	5
13	9..................	23　58.6	28　20	6.01	.06	5.138	6
14	B.A.C. 8366	23　59.4	29　18	5.84	.06	5.146	6
15	10	0　0.7	26　25	5.07	.02	5.136	6
16	β	0　3.3	31　27	2.32	0.11	2.835	2...3
17	O.A. 46	0　4.7	33　28	6.04	.08	5.146	6
18	12	0　18.7	28　47	5.57	.07	5.444	6
19	λ	0　25.7	36　5	4.93	.07	5.138	5
20	κ..................	0　26.9	27　40	4.32	.15	2.835	4...5
21	P. 0. 118	0　30.0	36　26	5.48	0.05	...	6
22	B.A.C. 148	0　30.2	30　17	5.79	.02	5.444	6
23	ζ..................	0　30.8	36　42	3.75	.05	2.835	4
24	α	0　34.3	34　4	2.41	.06	2.835	Var.
25	ξ..................	0　35.9	40　5	5.19	.16	...	6

Nos. 1 and 2. Argelander's place agrees better with the second of these stars.

No. 3. 5.22 is the mean of three accordant determinations depending on eighteen extinctions made on 1885.109, .442, and .448. H.P. = 4.09.

No. 6. 5.49 is the mean of three accordant determinations depending on eighteen extinctions made on same dates as No. 3. H.P. = 5.17.

No. 10. Birm., No. 652. The red colour is not salient.

No. 12. Σ 3049, Dist. 3″. Observed as one mass.

No. 21. 5.48 is the mean of three accordant determinations depending on eighteen extinctions made on 1885.146. .442, and .448. H.P. = 5.00.

No. 24. Birm., No. 9. The distant faint companion not observed. Variable from 2 2 mag. to 2.8 mag. Period irregular.

No. 25. 5.19 is the mean of three accordant observations, depending on eighteen extinctions made on 1882.835 and 1885.442, and .448. H.P. = 4.80.

Reference Number.	Star's Designation.	R.A. 1890.	N.P.D. 1890.	Adopted Zenithal Magnitude. Polaris 2.05.	Average Deviation in Magnitude.	Date, 1880 +	Mag. Argel. Uran.
		h. m.	° ′				
26	π	0 37.4	43 35	5.02	0.09	5.138	6
27	21	0 38.4	15 37	5.78	.01	5.138	6
28	B.A.C. 197	0 38.4	42 44	5.48	.08	5.146	6
29	ο	0 38.6	42 19	4.85	.09	5.442	5
30	P. o. 162	0 39.0	35 23	5.70	.04	...	6
31	23	0 40.4	15 45	5.54	0.06	5.442	6
32	η	0 42.3	32 46	3.41	.07	2.835	4...3
33	ν....................	0 42.6	39 38	4.93	.01	5.109	5
34	B.A.C. 228	0 44.0	26 21	5.65	.05	5.107	6
35	P. o. 209	0 46.5	29 29	5.13	.02	5.138	6
36	υ^1	0 48.5	31 37	5.07	0.03	5.107	6...5
37	γ....................	0 50.1	29 53	2.19	.09	2.810	2
38	υ^2	0 50.1	31 25	4.93	.02	5.107	6...5
39	B.A.C. 261........	0 51.5	24 15	6.01	.02	5.146	6
40	μ	1 0.7	35 36	5.40	.09	5.138	6
41	31	1 3.2	21 48	5.42	0.07	5.146	6
42	B.A.C. 335	1 4.3	26 23	5.41	.09	5.107	6...5
43	θ....................	1 4.4	35 25	4.68	.02	...	4..5
44	32	1 4.4	25 34	5.52	.03	5.146	6
45	φ	1 13.2	32 21	5.12	.04	5·146	5
46	ψ	1 18.1	22 27	5.02	0.08	5.107	5
47	δ....................	1 18.6	30 20	2.89	.06	2.810	3
48	38	1 23.0	20 18	5.92	.08	5.146	6
49	O.A. 1565	1 23.2	24 28	6.04	.04	5.146	6
50	χ	1 26.7	31 20	5.06	.06	5.107	6...5

No. 26. Variability suspected by Peirce. The recorded estimates differ.

No. 30. 5.70 is the mean of three accordant determinations, depending on eighteen extinctions made on 1885.130, .442, and .448. **H.P.** = 5.36.

No. 32. Σ 60, Binary. Observed as one object.

No. 37. Has double spectrum like T Coronæ. (Huggins.)

No. 40. Birm., Add. I, No. 6. The red colour is not salient.

No. 43. 4.68 is the mean of three accordant observations, depending on twenty-six extinctions made on 1882.210 and 1885.442, and .448. **H.P.** = 4.43.

No. 46. Σ 117, Dist. 32″. The faint distant companion not observed. The large star is bright yellow.

Reference Number.	Star's Designation.	R.A. 1890.	N.P.D. 1890.	Adopted Zenithal Magnitude. Polaris 2.05.	Average Deviation in Magnitude.	Date, 1880+	Mag. Argel. Uran.
		h. m.	° ′				
51	40	1 29.7	17 31	5.48	0.10	5.138	6
52	B.A.C. 482.........	1 30.9	32 36	5.85	.07	5.146	6
53	43	1 34.2	22 31	5.75	.03	5.442	6
54	42	1 34.4	19 56	5.20	.04	5.146	6
55	44	1 35.9	30 0	5.07	.06	5.138	6
56	ε	1 46.5	26 52	3.51	0.06	2.810	3...4
57	46	1 47.4	21 51	4.03	.05	5.107	5
58	B.A.C. 588.........	1 51.5	25 55	5.41	.04	5.146	6
59	48	1 52.9	19 38	4.77	.08	2.750	5...4
60	50	1 54.0	18 7	4.40	.03	...	4
61	47	1 54.1	13 15	5.37	0.05	5.146	5...6
62	52	1 54.7	25 38	5.86	.06	5.107	6
63	53	1 54.9	26 8	5.76	.02	5.146	6
64	49	1 55.4	14 25	5.51	.01	5.146	6...5
65	55	2 5.8	23 59	6.04	.02	5.107	6
66	ι	2 20.0	23 6	4.45	0.07	2.810	4
67	B.A.C. 777.........	2 27.6	17 40	5.54	.09	5.138	6...5
68	Bradley 417	3 0.0	16 1	4.79	.11	2.810	5...4

No. 51. Suspected to be variable in lustre. (See Nature, xxiii. 206.)
No. 60. 4.40 is the mean of three accordant determinations, depending on twenty-six extinctions, made on 1882.737 and 1885.442 and 1885.448. H.P. = 4.06.
No. 65. Ancient authorities differ in their recorded estimates of magnitude.
No. 66. Σ 262, Dist. 1″.8. Observed as one star.

CEPHEUS.

Reference Number.	Star's Designation.	R.A. 1890.	N.P.D. 1890.	Adopted Zenithal Magnitude. Polaris 2.05.	Average Deviation in Magnitude.	Date, 1880 +	Mag. Argel. Uran.
		h. m.	o ′				
1	κ.....................	20 12.6	12 37	4.50	0.04	2.901	4...5
2	θ.....................	20 27.7	27 23	4.24	.06	3.011	4
3	B.A.C. 7176	20 38.0	29 54	6.06	.02	5.084	6
4	B.A.C. 7193	20 40.3	29 48	6.16	.03	5.088	6
5	4.....................	20 41.8	23 45	5.50	.07	5.025	6...5
6	P. xx. 332	20 42.6	32 49	4.72	0.11	5.088	5...4
7	η	20 43.0	28 36	3.53	.05	3.011	4...3
8	D.M.+63°, No. 1663	20 47.4	26 23	6.27	.01	5.025	6
9	B.A.C. 7310	20 57.4	30 59	6.01	.05	5.088	6
10	R. 5091	20 59.1	33 46	5.98	.05	5.025	6
11	B.A.C. 7363	21 5.7	19 0	6.00	0.01	5.088	6
12	R. 5139	21 7.1	27 9	6.33	.04	5.084	6
13	B.A.C. 7381	21 7.6	12 19	5.55	.12	5.084	6
14	B.A.C. 7377	21 9.0	30 28	5.70	.03	5.025	6...5
15	α	21 15.9	27 53	2.57	.04	2 895	3...2
16	B.A.C. 7417	21 16.2	31 50	5.84	0.05	5.088	6
17	6.....................	21 17.1	25 36	5.40	.06	5.025	6
18	B.A.C. 7430	21 17.6	29 42	6.52	.04	5.382	6
19	7.....................	21 25.6	23 40	5.58	.02	5.089	6
20	β	21 27.2	19 55	3.37	.05	2·895	3
21	B.A.C. 7495	21 28.0	30 2	5.41	0.14	5.089	6
22	9.....................	21 35·0	28 25	4.99	.07	5.382	5...6
23	P. xxi. 248	21 35.5	33 1	5.67	.05	5.025	6...5
24	μ	21 40.1	31 44	4.53	.18	3.011	6...5
25	11	21 40.3	19 12	4.92	.09	5.084	5

No. 1. Σ 2675, Dist. 7″. Observed as one star.
No. 9. The B.A.C. assigns this star to Cygnus.
No. 10. Σ 2751, Dist. 1″.9. Observed as one mass. Probably slow binary.
Nos. 11 and 14. The B.A.C. assigns these stars to Draco.
No. 14. Σ 2780, Dist. 1″. Observed as one star.
No. 16. Σ 2790, Dist. 4″.5. Colour orange. Observed as one mass.
No. 18. Difficulty in identifying Argelander's star.
No. 20. Σ 2806, Dist. 14″. The faint companion not observed.
No. 23. Σ 2816, Dist. 120″. The companion not observed.
No. 24. Birm., No. 594. The orange colour is very salient. Variable from 4 to 5 magnitude.
 The period is irregular.

Reference Number.	Star's Designation.	R.A. 1890.	N.P.D. 1890.	Adopted Zenithal Magnitude. Polaris 2.05.	Average Deviation in Magnitude.	Date, 1880 +	Mag. Argel. Uran.
		h. m.	o ′				
26	P. xxi. 302............	21 41.7	18 11	5.53	0.03	5.094	6
27	R. 5390	21 41.9	28 3	6.23	.08	5.044	6
28	ν..........	21 42.3	29 23	4.84	.07	...	5
29	B.A.C. 7658	21 53.6	26 54	5.36	.08	5.382	6...5
30	16	21 57.7	17 21	5.38	.01	5.094	5...6
31	14	21 58.4	32 32	5.60	0.10	5.100	6
32	15	22 0.3	30 43	6.41	.07	...	6
33	ξ.................	22 0.6	25 55	4.72	.05	2.895	5...4
34	20	22 1.7	27 45	5.65	.04	5.044	6
35	19	22 1.8	28 15	5.38	.01	5.094	6...5
36	ζ.................	22 7.0	32 21	3.39	0.04	3.011	4...3
37	24	22 7.7	18 12	4.74	.07	5.094	5...4
38	λ	22 7.8	31 8	5.50	.06	5.044	6...5
39	B.A.C. 7754	22 7.8	33 43	5.58	.16	5.094	6
40	B.A.C. 7760	22 8.2	20 25	5.60	.07	5.382	6
41	B.A.C. 7759	22 8.4	29 48	5.59	0.09	5.382	6
42	B.A.C. 7766	22 8.9	27 15	6.06	.11	5.044	6
43	ε.................	22 11.0	33 30	4.70	.05	...	5...4
44	25	22 14.6	27 45	6.28	.03	5.094	6
45	P. xxii. 165	22 22.2	4 27	5.50	.01	5.103	5...6
46	O.A. 4148	22 23.2	19 48	5.59	0.14	5.100	6
47	26	22 23.5	25 26	5.73	.12	5.044	6...5
48	δ.................	22 25.1	32 9	4.21	.08	3.011	Var.
49	ρ {28	22 25.9	11 46	5.82	.03	5.103	6...5
50	{29	22 28.9	11 44	5.43	.06	5.382	6...5

No. 26. The B.A.C. assigns this star to Draco.

No. 28. 4.84 is the mean of three accordant determinations, depending on twenty-six extinctions, made 1885.044, .382, and .437. H.P. = 4.50.

No. 29. Birm., No. 599. The red colour is not salient.

No. 32. 6.41 is the mean of two measures (6.40, 6.42) observed with six extinctions on 1885.044, and with also six extinctions on 1885.382.

No. 33. Σ 2863, Dist. 6″. Observed as one star. This star was re-observed with six extinctions 1885.382, the resulting magnitude was 4.60. H.P. = 4.45.

No. 34. Birm., No. 602. The colour of the star was noted to be orange.

No. 36. Birm., No. 604. The colour is not salient.

No. 40. Σ 2883, Dist. 15″. The companion was not observed.

No. 42. Birm., No. 606. The red colour is not salient.

No. 43. 4.70 is the mean of two determinations (4.78, 4.74) observed 1883.011 with twenty extinctions, and 1885.382 with six extinctions. H.P. = 4.24.

No. 45. B.A.C. assigns this star to Ursa Minor.

No. 48. Σ 58¹, Dist. 40″. The companion was not observed. Variable from mag. 3.7 to 4.9. Period about 5ᵈ, 8ʰ 30ᵐ.

Reference Number.	Star's Designation.	R.A. 1890.		N.P.D. 1890.		Adopted Zenithal Magnitude. Polaris 2.05.	Average Deviation in Magnitude.	Date, 1880+	Mag. Argel. Uran.
		h. m.		o '					
51	B.A.C. 7876	22	29.8	20	39	6.11	0.14	5.044	6
52	B.A.C. 7881	22	30.4	14	20	5.65	.08	5.100	5...6
53	31	22	33.0	16	56	5.30	.09	5.044	5
54	30	22	34.8	26	59	5.45	.02	5.044	5...6
55	ι	22	45.8	24	23	3.61	.07	...	4...3
56	P. xxii. 258	22	47.9	7	26	5.07	0.07	5.100	5
57	P. xxii. 295	22	55.3	6	14	5.16	.05	5.094	5...6
58	B.A.C. 8039	22	59.4	23	22	5.65	.01	5.044	6
59	π	23	4.4	15	13	4.67	.11	2.841	5...4
60	B.A.C. 8104	23	10.7	16	22	5.73	.11	5.044	6
61	B.A.C. 8106	23	11.4	19	43	5.61	0.08	5.103	6
62	o	23	14.1	22	29	5.22	.08	5.044	6...5
63	B.A.C. 8180	23	22.6	20	15	5.98	.06	...	6
64	P. xxiii. 135	23	27.8	3	18	5.74	.17	5.103	6
65	B.A.C. 8217	23	30.2	18	58	6.34	.00	5.094	6
66	γ	23	34.8	12	59	3.51	0.09	2.841	3...4
67	P. xxiii. 191	23	42.6	22	48	5.17	.01	5.044	6
68	B.A.C. 8321	23	51.2	7	25	6.19	.07	5.103	6
69	B.A.C. 154	0	31.5	8	7	6.30	.14	5.044	6
70	B.A.C. 225	0	44.5	6	53	5.70	.04	5.094	6
71	P. o. 220	0	53.5	4	20	4.66	0.08	2.841	4...5
72	P. o. 283	1	2.7	10	56	5.70	.06	5.044	6...5
73	B.A.C. 393	1	14.1	11	51	5.97	.00	5.044	6
74	B.A.C. 605	1	55.9	9	14	6.00	.04	5.103	6
75	B.A.C. 784	2	31.9	9	1	6.04	.05	5.044	6

No. 55. 3.61 is the mean of two measures (3.59, 3.62) observed 1882.841 and 1882.895, each with twenty extinctions.

No. 59. OΣ 589, Dist. 1".3. Binary. Observed as one star.

No. 62. Σ 3001, Dist. 2".5. Binary. Observed as one star.

No. 63. 5.98 is the mean of two measures (6.02, 5.94) observed with six extinctions on 1885.092, and on 1885.382. H.P. = 5.71.

No. 64. B.A.C. assigns this star to Ursa Minor.

No. 71. Birm., No. 12. The red colour is not salient. This star is called 2 Ursæ Minoris in the B.A.C.

Nos. 72, 73, 74, and 75. The B.A.C. assigns these stars to Cassiopeia.

Cepheus.

Reference Number.	Star's Designation.	R.A. 1890.	N.P.D. 1890.	Adopted Zenithal Magnitude, Polaris 2.05.	Average Deviation in Magnitude.	Date, 1880+	Mag. Argel. Uran.
		h. m.	° ′				
76	P. ii. 91	2 51.4	11 1	5.89	0.05	5.094	6
77	B.A.C. 908	2 54.6	8 57	5.02	.06	5.044	6
78	B.A.C. 960	3 6.2	5 29	6.05	.04	5.094	6
79	P. iii. 255	3 6.3	12 40	5.54	.10	5.103	6
80	B.A.C. 1061	3 30.4	3 42	5.93	.07	5.386	6
81	P. iii. 160	3 51.6	9 36	5.46	0.13	5.094	5...6
82	B.A.C. 1247	4 2.8	6 27	5.55	.04	5.044	6
83	B.A.C. 1263	4 5.9	6 56	5.59	.04	5.103	6
84	B.A.C. 1276	4 7.9	9 26	5.01	.05	5.100	6
85	B.A.C. 1448	4 39.8	8 59	5.20	.11	5.100	5
86	R. 1311	4 52.9	4 10	6.45	0.12	5.103	} 6
87	R. 1377	5 6.5	4 24	6.10	.07	5.103	
88	B.A.C. 1662	5 29.8	4 51	6.51	.06	5.100	6
89	P. vi. 21	6 48.8	2 47	5.44	.08	5.103	5

Nos. 76, 77, 79, and 81. The B.A.C. assigns these stars to Cassiopeia.
No. 76. Σ 320, Dist. 5″. Observed as one mass.
No. 77. Suspicion of variability. (See Nature, xxiii. 206 and xxvii. 541.)
Nos. 78, 80, 82, 83, 88, and 89. B.A.C. assigns this star to Ursa Minor.
No. 79. Suspected by Struve to be variable. (Berl. Jahr. 1819, p. 186.)
No. 81. Σ 460, Dist. <1″. Observed as one star.
Nos. 84 and 85. B.A.C. assigns this star to Camelopardalus.
Nos. 86 and 87. It is uncertain which of these is the star described by Argelander. The position recorded in the Uranometria agrees more nearly with the second and brighter star.
No. 89. Birm., No. 154. The red colour is not salient.

CETUS.

Reference Number.	Star's Designation.	R.A. 1890.	N.P.D. 1890.	Adopted Zenithal Magnitude. Polaris 2.05.	Average Deviation in Magnitude.	Date, 1880+	Mag. Argel. Uran.
		h. m.	° ′				
1	2...................	23 58.1	107 57	...	0.08	2.742	4...5
2	4...................	0 2.1	93 10	6.52	.10	3.886	} 6
3	5...................	0 2.6	93 4	6.59	.07	3.886	
4	P. o. 1	0 4.7	95 52	6.15	.07	4.772	6
5	6...................	0 5.7	106 4	...	.15	2.742	5...4
6	L. 158	0 8.8	98 24	5.47	0.05	3.886	6
7	7	0 9.1	109 33	...	.02	2.742	5...4
8	ι...................	0 13.8	99 26	3.66	.08	2.742	3...4
9	13	0 29.6	94 12	5.77	.05	4.776	6...5
10	P. o. 146	0 35.1	94 57	6.25	.08	4.776	6
11	β	0 38.1	108 35	...	0.10	2.742	2
12	20	0 47.4	91 44	5.22	.06	3.871	5...6
13	25	0 57.5	95 25	6.05	.04	4.843	6
14	η...................	1 3.1	100 46	3.47	.05	2.827	3
15	37	1 8.9	98 31	5.19	.07	4.843	6
16	38	1 9.2	91 34	6.02	0.06	3.886	6
17	39	1 11.0	93 5	5.77	.04	3.866	6
18	42	1 14.2	91 5	6.46	.08	4.843	6
19	θ	1 18.5	98 45	3.36	.07	...	3
20	W.B. i. 271	1 18.8	97 29	6.05	.04	3.871	6

The magnitude, corrected for atmospheric absorption, for the stars not given in the text is as follows :—

No. 1 .... 4.34.		No. 25 .... 4.70.	
No. 5 .... 4.63.		No. 26 .... 3.47.	
No. 7 .... 4.32.		No. 27 .... 3.57.	
No. 11 .... 2.42.		No. 47 .... 4.51.	
No. 23 .... 3.13.		No. 50 .... 3.07.	

Nos. 8 and 11. Birm., Nos. 3 and 10. The red colour is not salient.

No. 12. Spectrum II a ! ! (Vogel.) Called red in the Uranometria Argentina.

No. 14. Birm., No. 16. Noted to be of orange tint.

No. 15. Σ 3¹, Dist. 50″. The companion not observed.

No. 18. Σ 113, Dist. 1″.2. Observed as one mass.

No. 19. 3.36 is the mean of four determinations made on as many nights at Cairo, between 1883.099 and 1883.107. H.P. = 3.77.

Reference Number.	Star's Designation.	R.A. 1890.	N.P.D. 1890.	Adopted Zenithal Magnitude. Polaris 2.05.	Average Deviation in Magnitude.	Date. 1880 +	Mag. Argel. Urau.
		h. m.	° ′				
21	L. 2798	1 28.2	97 34	6.17	0.06	3.902	6
22	L. 3159	1 38·2	94 14	5.63	.06	3.917	5...6
23	τ	1 39.0	106 32	...	.04	2.827	3...4
24	P. i. 167	1 40.5	96 17	5.90	.05	4.792	6
25	χ	1 44.2	101 14	...	.10	2.827	5...4
26	ζ	1 46.0	100 53		0.13	2.827	3
27	υ	1 54.8	111 37	...	.07	2.827	4
28	L. 3717	1 55.0	99 3	5.95	.06	3.917	} 6
29	L. 3731	1 55.4	99 0	7.40	.09	3.917	
30	60	1 57.5	90 24	5.68	.08	4.792	6
31	64	2 5.6	81 57	5.88	0.09	3.924	6
32	63	2 6.0	92 20	6.22	.05	3.917	6
33	ξ¹	2 7.2	81 40	4.51	.08	2.827	4...5
34	67	2 11.5	96 55	5.84	.05	3.924	6
35	P. ii. 52	2 12.3	88 46	6.02	.06	4.792	6
36	69	2 16.3	90 6	5.91	0.05	4.792	6
37	70	2 16.6	91 23	5.86	.04	4.843	6
38	ξ²	2 22.3	82 2	4.66	.10	2.827	4
39	B.A.C. 776	2 25.8	88 13	5.62	.04	3.871	6
40	75	2 26.6	91 31	5.78	.05	3.871	6...5
41	P. ii. 123	2 30.0	83 39	6.06	0.03	4.843	6
42	ν.	2 30.1	84 53	4.93	.06	2.827	5
43	80	2 30.6	98 18	5.84	.04	4.792	6
44	P. ii. 130	2 30.8	82 45	6.17	.06	3.917	6
45	81	2 32.2	93 52	5.68	.05	3.924	6
46	δ	2 33.8	90 9	4.23	0.08	...	4
47	ε	2 34.2	102 20	...	.09	2.832	5...4
48	P. ii. 148	2 34.5	84 22	6.38	·10	4.792	6
49	γ	2 37.6	87 14	3.38	.08	2·832	3...4
50	π	2 38.9	104 19		.07	2.832	4

No. 26. Probably variable. (See Uranometria Argentina, p. 312.)
Nos. 30 and 38. Spectrum I a ! (Vogel.)
No. 36. Spectrum III a ! ! (Vogel.)
No. 39. Spectrum II a ! (Vogel.)
No. 42. Σ 281, Dist. 6″. Observed as one mass.
No. 46. 4.23 is the mean of two measures (4.21, 4.37) after applying the correction 0.06 for atmospheric absorption, observed 1882.827 and 1882.832, each with twenty extinctions.
No. 49. Σ 299, Dist. 2′.5. Observed as one mass. Spectrum I a ! (Vogel.)
No. 50. Ancient estimates discordant. There is probability of slight variability in this star.

Refer-ence Number.	Star's Designation.	R.A. 1890.	N.P.D. 1890.	Adopted Zenithal Magnitude. Polaris 2.05.	Average Deviation in Magnitude.	Date, 1880+	Mag. Argel. Uran.
		h. m.	° ′				
51	μ	2 39.0	80 21	4.25	0.06	2·832	4
52	λ	2 53.8	81 32	4.93	.06	...	5...4
53	a....................	2 56.5	86 21	2.44	.05	2.832	2...3
54	94	3 7.2	91 36	5.30	.05	4.792	5...6
55	95	3 12.7	91 20	5.77	.05	4.843	6
56	κ	3 13.6	87 2	5.15	0.09	3.924	5
57	97	3 15.4	86 43	5.71	.04	4.792	6

No. 51. B.A.C. assigns this star to Aries. Spectrum I a ! (Vogel.)

No. 52. 4.93 is the mean of two measures (4.96, 4.90) observed 1882.832 with twenty extinctions, and on 1884.967 with six. H.P. = 4.60. Spectrum I a ! (Vogel.)

No. 53. Spectrum III a ! ! ! (Vogel.) By Huggins, the Spectrum is nearly the same as that of α Orionis.

COMA.

1	2	11 58.6	67 56	5.90	0.02	5.333	6
2	4	12 6.3	63 31	6.06	.06	4.386	6
3	5	12 6.6	68 51	5.43	.07	4.357	6
4	6	12 10.4	74 29	5.15	.02	4.357	5
5	7	12 10.8	65 27	5.36	.05	4.385	5...6
6	W.B. xii. 199 ...	12 12.0	60 26	5.56	0.05	4.386	6
7	11	12 15.2	71 36	4.73	.06	4.357	5
8	P. xii. 57	12 16.5	64 37	6.05	.04	4.385	6
9	12	12 17.0	63 33	4.64	.03	4.386	5
10	13	12 18.8	63 17	5.47	.05	...	5
11	14	12 20.9	62 7	5.09	.06	4.385	5...4
12	γ	12 21.6	61 7	4.55	.08	5.333	4...5
13	16	12 21.5	62 34	5.39	.06	...	5
14	17	12 23.4	63 29	5.07	.08	4.386	5
15	18	12 24.0	65 17	5.41	.03	4.385	6

No. 2. Σ 1596, Dist. 3″.6. Observed as one mass.

No. 4. Spectrum I a ! (Vogel.)

No. 9. Has a distant companion, not observed.

No. 10. 5.47 is the mean of two measures (5.43, 5.51) observed with twenty extinctions 1884.365, and with six on 1885.333. H.P. = 5.09.

No. 12. Birm., No. 279. The colour is yellow.

No. 13. 5.39 is the mean of two measures (5.36, 5.42) observed with twenty extinctions 1884.386, and with six 1885.333. H.P. = 5.00.

Reference Number.	Star's Designation.	R.A. 1890.	N.P.D. 1890.	Adopted Zenithal Magnitude. Polaris 2.05.	Average Deviation in Magnitude.	Date, 1880 +	Mag. Argel. Uran.
		h. m.	° ′				
16	20	12 24.2	68 30	5.85	0.02	4.357	6
17	21	12 25.5	64 50	5.60	.06	4.385	6...5
18	23	12 29.4	66 46	4.95	.05	4.357	5
19	24	12 29 6	71 1	4.79	.06	4.386	5
20	26	12 33.7	68 20	5.47	.06	4.357	6
21	27	12 41.2	72 49	5.23	0.06	4.357	5
22	29	12 43.4	75 17	5.96	.02	4.385	6
23	30	12 43.9	61 51	6.24	.05	4.357	6
24	31	12 46.4	61 52	5.14	.03	4.386	5
25	35	12 47.9	68 9	4.96	.05	4.385	5
26	36	12 53.5	72 0	5.22	0.05	4.357	5
27	37	12 55.0	58 37	5.00	.04	4.385	5
28	39	13 1.0	68 15	6.14	.07	4.357	6
29	40	13 1.0	66 48	5.98	.08	4.386	6
30	41	13 1.9	61 47	5.00	.05	4.385	5
31	α	13 4.7	71 53	4.38	0.05	2.452	4...5
32	β	13 6.8	61 34	4.24	.08	2.452	4
33	P. xiii. 18	13 7.2	70 40	6.68	.05	4.357	} 6
34	W.B. xiii. 98 ...	13 7.9	70 42	6.47	.04	4.357	
35	P. xiii. 36	13 11.2	69 38	6.33	.03	4.385	6
36	P. xiii. 77	13 19.9	65 34	5.56	0.04	...	6
37	W.B. xiii. 596 ...	13 31.8	64 50	6.04	.03	4.386	6

No. 19. Σ 1657, Dist. 20″. The distant companion not observed.

No. 22. = 36 Virginis. (See Introduction to B.A.C., p. 75.)

No. 25. Σ 1687, Dist. 1″. Triple. Observed as one mass. The third distant star not observed.

No. 26. Birm., No. 300. The red colour is not salient. Spectrum III a ! (Vogel.) Variable?

No. 27. = 13 Canum Ven. (See Introduction to B.A.C., p. 75, where however the star is incorrectly called 31 Comæ.)

No. 31. Σ 1728, Dist. < 1″. Binary. Period 26 years. Observed as one mass.

No. 36. 5.56 is the mean of two measures (5.54, 5.57) observed 1884.357 with twenty extinctions, and on 1885.333 with six. H.P. = 5.93.

No. 37. Birm., No. 311. The colour is not salient.

CORONA.

Reference Number.	Star's Designation.	R.A. 1890.	N.P.D. 1890.	Adopted Zenithal Magnitude. Polaris 2.05.	Average Deviation in Magnitude.	Date, 1880 +	Mag. Argel. Uran.
		h. m.	o ′				
1	o..................	15 15.6	59 59	5.61	0.06	4.530	6
2	η	15 18.7	59 19	5.05	.03	4.535	5
3	β	15 23.3	60 31	3.85	.09	2.485	4...3
4	θ..................	15 28.5	58 16	4.26	.07	2.583	4
5	a..................	15 30.0	62 55	2.23	.10	...	2
6	μ	15 31.2	50 38	5.44	0.04	4.538	5
7	W.B. xv. 717......	15 32.4	59 39	6.34	.04	4.530	6
8	P. xv. 142	15 33.6	65 7	6.99	.09	4.538	6
9	ζ..................	15 35.2	53 0	4.66	.05	5.352	4
10	γ..................	15 38.1	63 21	3.91	.08	...	4...3
11	π	15 39.7	57 8	5.76	0.06	4.530	6
12	δ..................	15 45.0	63 36	4.87	.05	...	4...5
13	κ..................	15 47.1	54 0	5.02	.09	2.583	5...4
14	λ	15 51.8	51 44	5.79	.05	5.352	6...5
15	ε..................	15 53.0	62 48	4.18	.04	2.583	4
16	W.B. xv. 239......	15 54.9	53 3	5.81	0.08	4.532	6
17	ρ..................	15 56.9	56 21	5.08	.06	4.530	6...5
18	ι	15 57.0	59 51	5.07	.09	2.583	5...4
19	P. xv. 266	15 59.3	53 4	6.08	.05	...	6
20	τ..................	16 5.0	53 14	5.07	.09	2.583	5...4

No. 2. Σ 1937, Dist. 1″. Binary. Period 40 years. Observed as one mass.

No. 5. Birm., No. 354. The red colour is not salient. Seidel suggests variability. (Resul. Phot. Mess., p. 163.)

No. 5. 2.23 is the mean of three accordant determinations made 1882.485, 1883.198, and 1883.201.

No. 9. Σ 1965, Dist. 6″. Observed as one mass. Birm. Add. II, No. 12. Colour not salient.

No. 10. Σ 1967, Dist. <1″. Period 96 years. Observed as one mass. 3.91 is the mean of two measures (3.87, 3.95) observed 1882.583 with twenty extinctions, and 1885.352 with six. H.P. = 4.18.

No. 12. Birm., Add. II, No. 13. The red colour is not salient. 4.87 is the mean of two measures (4.93 and 4.80) observed 1882.430 with twenty extinctions, and 1885.352 with six. H.P. = 4.56.

No. 15. Birm., Add. II, No. 14. The star is slightly orange coloured.

Reference Number.	Star's Designation.	R.A. 1890.	N.P.D. 1890.	Adopted Zenithal Magnitude. Polaris 2.05.	Average Deviation in Magnitude.	Date, 1880+	Mag. Argel. Uran.
		h. m.	° ′				
21	P. xvi. 25	16 7.8	53 17	5.92	0.07	5.352	6
22	σ	16 10.6	55 52	5.13	.07	4.538	6
23	υ.................	16 12.3	60 35	5.61	.03	4.535	6...5
24	ξ.................	16 17.8	58 51	4.63	.06	4.530	5
25	ν¹	16 18.2	55 57	5.07	.02	4.535	} 5
26	ν²	16 18.3	56 3	5.08	0.01	4.535	

No. 22. Σ 2032, Dist. 1″.3. Binary. Observed as one mass.
Nos. 25 and 26. These stars form Σ 29¹. They are also in Birmingham's Catalogue, Nos. 376 and 377. The stars are yellow.

CYGNUS.

1	B.A.C. 6579	19 9.3	40 22	5.92	0.05	4.932	6
2	O.A. 19116	19 13.7	43 12	5.93	.06	4.938	6
3	κ	19 14.6	36 50	3.83	.04	2.455	4
4	2.................	19 19.8	60 36	5.10	.02	4.888	5
5	4.................	19 22.2	53 54	5.08	.09	4.888	5
6	W.B. xix. 667 ...	19 22.7	45 13	6.50	0.03	4.927	6
7	7..................	19 24.8	37 54	5.95	.04	4.938	6
8	β¹	19 26.3	62 16	3.02	.10	2.455	} 3
9	β²	19 26.3	62 16	4.83	.07	2.465	
10	W.B. xix. 779 ...	19 26.7	54 0	6.04	.04	4.899	6
11	ι.................	19 26.9	38 30	4.02	0.05	2.430	4
12	8..................	19 27.7	55 47	4.75	.03	2.455	5...4
13	O.A. 19337	19 28.5	39 56	5.91	.06	4.932	6
14	9.................	19 30.4	60 47	5.62	.05	4.888	6
15	B.A.C. 6718	19 31.1	47 50	5.37	.09	5.415	6
16	P. xix. 211........	19 31.5	39 0	5.60	0.09	4.938	6
17	11	19 31.9	53 18	5.90	.07	4.888	6
18	B.A.C. 6731	19 33.2	45 33	5.44	.05	4.910	6
19	θ	19 33.5	40 2	4.00	.03	2.430	5...4
20	φ	19 35.0	60 6	5.11	.06	4.888	5

No. 1. Σ 2486, Dist. 10″. Observed as one mass.
Nos. 8 and 9. These stars form Σ 43¹, Dist. 34″, also Birm., No. 503. There is strong suspicion of variability. (See Ast. Nach., lxx. p. 108.)

Refer-ence Number.	Star's Designation.	R.A. 1890.	N.P.D. 1890.	Adopted Zenithal Magnitude. Polaris 2.05.	Average Deviation in Magnitude.	Date, 1880+	Mag. Argel. Uran.
		h. m.	° ′				
21	14	19 35.8	47 26	5.47	0.12	4.910	6...5
22	B.A.C. 6748	19 36.2	35 17	5.69	.02	4.938	6
23	B.A.C. 6754	19 37.4	44 44	5.27	.15	4.910	6...5
24	R. 4420	19 38.2	50 0	6.02	.08	4.899	6
25	L. 37527	19 38.4	57 49	6.07	.05	4.888	6
26	16	19 38.9	39 44	6.11	0.05	4.927	} 6...5
27	P. xix. 262.........	19 39.0	39 45	6.12	.03	4.927	
28	15	19 40.3	52 55	5.42	.05	...	5...6
29	δ	19 41.5	45 8	2.79	.05	...	3
30	17	19 42.3	56 31	5.31	.05	4.888	5...6
31	B.A.C. 6799	19 44.2	42 22	6.30	0.07	4.932	6
32	19	19 46.7	51 34	5.68	.04	4.899	6
33	B.A.C. 6817	19 46.8	49 41	5.76	.05	4.910	6
34	20	19 47.9	37 17	5.27	.03	5.415	5...6
35	O.A. 19720	19 48.7	43 15	5.74	.03	4.927	6
36	B.A.C. 6830	19 48.9	42 21	5.96	0.05	4.932	6
37	23	19 51.0	32 46	5.34	.03	4.899	5...6
38	22	19 51.9	51 48	5.14	.05	...	5...6
39	B.A.C. 6852	19 51.6	30 35	6.02	.04	4.941	6...5
40	η	19 52.2	55 10	4.10	.07	2.430	4...5
41	ψ	19 52.8	37 51	4.79	0.04	4.938	5
42	B.A.C. 6857	19 53.4	49 56	5.74	.05	4.899	6...5
43	B.A.C. 6865	19 53.7	39 24	6.05	.04	4.938	6
44	B.A.C. 6867	19 53.8	31 27	5.37	.05	4.941	6...5
45	O.A. 19835	19 54.1	33 36	5.89	.10	4.938	6

Nos. 26 and 27. These stars form Σ 46¹, Dist. 37″. The two stars, seen as one object, are called 16, *c*, in Argelander.

No. 28. 5.42 is the mean of two measures (5.44, 5.39) observed 1884.899 and 1885.415, each with six extinctions. H.P. = 5.03.

No. 29. Σ 2579, Dist. 1″.5. Binary. Observed as one mass. 2.79 is the mean of two measures (2.79, 2.78) observed 1882.430 with twenty extinctions, and 1885.415 with six.

No. 30. Called χ in the B.A.C. Σ 2580, Dist. 26″. The brighter star only observed.

No. 32. Birm., No. 519. The red colour is not salient.

No. 38. 5.14 is the mean of two measures (5.15, 5.13) observed 1884.910 and 1885.415, each with six extinctions. H.P. = 4.67.

No. 39. B.A.C. assigns this star to Draco.

No. 40. Possibly variable. (See Bonn. Beob., vol. vii, p. 402.)

No. 41. Σ 2605, Dist. 3″.5. Observed as one mass.

No. 44. B.A.C. assigns this star to Cepheus.

Reference Number.	Star's Designation.	R.A. 1890.	N.P.D. 1890.	Adopted Zenithal Magnitude. Polaris 2.05.	Average Deviation in Magnitude.	Date, 1880+	Mag. Argel. Uran.
		h. m.	° ′				
46	W.B. xix. 1739 ...	19 54.2	59 19	5.55	0.09	5.415	6
47	L. 38193	19 54.2	48 2	6.46	.05	4.910	6
48	25	19 55.8	53 16	5.52	.07	...	6...5
49	B.A.C. 6876	19 55.9	44 32	5.96	.03	4.927	6
50	P. xix. 380	19 56.4	38 15	6.05	.04	4.932	6
51	26	19 58.3	40 12	5.34	0.06	4.927	6...5
52	W.B. xix. 1910 ...	19 59.0	60 23	5.74	.05	4.888	6
53	W.B. xix. 1957 ...	20 0.2	58 5	5.74	.05	4.899	6
54	O.A. 19983	20 1.2	42 5	6.04	.02	4.927	6
55	B.A.C. 6918	20 2.1	38 28	6.13	.02	4.932	6
56	27	20 2.3	54 20	5.58	0.07	4.899	6...5
57	B.A.C. 6928	20 3.3	37 10	5.89	.06	4.941	6...5
58	28	20 5.3	53 29	4.97	.09	4.899	5
59	B.A.C. 6959	20 9.5	38 52	6.42	.03	4.932	6
60	c^1_1 30	20 9.8	43 31	4.73	.09	2.430	4
61	c^1_2 31.............	20 10.2	43 36	4.25	0.10	2.430	
62	29	20 10.4	53 32	4.99	.07	4.899	5
63	33	20 10.8	33 46	4.47	.09	5.415	4...5
64	L. 38943	20 11.1	56 36	5.97	.08	4.888	6
65	c^2 32	20 12.1	42 37	4.46	.06	4.941	4...5
66	O.A. 20293	20 12.4	44 45	6.03	0.10	4.910	6
67	B.A.C. 6986	20 13.0	49 58	5.58	.03	4.910	6
68	36	20 14.3	53 21	5.09	.03	4.927	6
69	35	20 14.4	55 22	5.34	.07	4.927	5...6
70	R. 4734	20 15.8	34 57	5.81	.05	4.941	6...5

No. 47. Σ 2607, Dist. 3″.5. Observed as one mass.

No. 48. 5.52 is the mean of two measures (5.47, 5.57) observed 1884.899 and 1885.415, each with six extinctions. H.P. = 5.20.

No. 51. (Σ unnum.) The distant companion not observed.

Nos. 60 and 61. Seen as one star by Argelander.

No. 61. Birm., No. 543. The colour is bright yellow.

No. 63. The observation of 1882.430 (see Memoirs of R.A.S., vol. xlvii, p. 431) is rejected. The night was hazy, and with cumulus clouds in parts of the sky.

Nos. 65 and 67. Birm., Nos. 549 and 551. The colour of both is orange.

No. 70. Σ 2671, Dist. 3″. Observed as one mass.

Reference Number.	Star's Designation.	R.A. 1890.	N.P.D. 1890.	Adopted Zenithal Magnitude. Polaris 2.05.	Average Deviation in Magnitude.	Date, 1880 +	Mag. Argel. Uran.
		h. m.	° ′				
71	R. 4738	20 16.4	43 30	6.10	0.04	4.932	6
72	γ	20 18.3	50 6	2.26	.04	...	3 . . 2
73	O.A. 20430	20 18·5	44 34	5.96	.04	4.941	6
74	B.A.C. 7027	20 18.8	49 20	6.10	.04	4.912	6
75	39	20 19.5	58 10	4.81	.07	4.927	5
76	W.B. xx. 665......	20 19.6	52 53	5.95	0.05	4.912	6
77	40	20 23.5	51 56	5.76	.05	4.912	6
78	B.A.C. 7064	20 23.7	33 43	6.29	.02	4.946	6
79	41	20 24.9	60 0	4.32	.06	2.455	4...5
80	42	20 25.1	53 55	6.05	.01	4.912	6
81	ω² (45)	20 26.7	41 25	5.04	0.05	4.927	5
82	B.A.C. 7086	20 26.8	34 18	6.07	.03	4.946	6
83	ω³ (46)	20 27.9	41 9	5.64	.04	4.932	6 . . 5
84	B.A.C. 7105	20 29.1	33 36	6.16	.03	4.946	6
85	47	20 29.6	55 8	5.03	.03	...	5...6
86	B.A.C. 7112	20 30.3	43 41	5.74	0.02	4.927	6
87	48	20 33.0	58 49	5.92	.09	4.888	} 6
88	P. xx. 243	20 33.1	58 52	6.12	.06	4.888	
89	L. 39885	20 33.2	52 3	6.31	.06	4.912	6
90	B.A.C. 7158	20 35.5	49 49	6.05	.03	4.927	6
91	W.B. xx. 1193 ...	20 36.1	46 55	6.20	0.02	4.927	6
92	49	20 36.6	58 5	5.93	.05	4.912	6
93	α	20 37.7	45 8	1.32	.04	...	2...1
94	P. xx. 283	20 38.0	54 57	6.30	.07	4.912	6
95	B.A.C. 7174	20 38.0	48 41	5.71	.02	4.927	6

No. 72. **2.26** is the mean of two determinations (2.28, 2.23) made on 1882.430 with twenty extinctions, and on 1883.070 with ten.

No. 74. Birm., No. 555. The red colour is not salient.

No. 75. Birm., No. 557. The colour is yellow.

No. 78. B.A.C. assigns this star to Cepheus. Σ 2687, Dist. 26″. The brighter star only observed.

Nos. 82 and 84. B.A.C. assigns these stars to Cepheus.

No. 83. Σ Unnum., Dist. 55″. The distant companion not observed. Birm., No. 562. The red colour is not salient.

No. 85. **5.03** is the mean of two determinations (5.09, 4.98) made on 1884.912 and 1885.415, each with six extinctions. H.P. = 4.77.

Nos. 87 and 88. These stars form Σ 53¹, Dist. 178″.

No. 92. Σ 2716, Dist. 2″.7. Observed as one mass.

No. 93. **1.32** is the mean of six accordant determinations, made on as many nights, three at Oxford and three at Cairo, involving seventy extinctions, made between 1882.430 and 1883.170.

Reference Number.	Star's Designation.	R.A. 1890.	N.P.D. 1890.	Adopted Zenithal Magnitude. Polaris 2.05.	Average Deviation in Magnitude.	Date, 1880+	Mag. Argel. Uran.
		h. m.	° ′				
96	51	20 38.8	40 3	5.53	0.06	4.946	6...5
97	W.B. xx. 1276 ...	20 39.0	54 48	6.26	.07	4.912	6
98	B.A.C. 7198	20 41.0	43 6	6.55	.03	4.932	6
99	52	20 41.0	59 41	4.53	.03	2.455	4..5
100	ε	20 41.8	56 27	2.45	.07	2.455	3...2
101	λ	20 43.1	53 54	4.90	0.06		5...4
102	55	20 45.2	44 17	5.39	.06	...	6...5
103	56	20 46.2	46 21	5.25	.03	4.938	5...6
104	57	20 49.3	46 2	4.84	.02	5.415	5..6
105	B.A.C. 7254	20 49.5	45 14	5.79	.05	4.941	6
106	B.A.C. 7268	20 51.4	43 0	6.00	0.09	4.956	6
107	ν	20 53.1	49 15	4.23	.06	2.455	4
108	B.A.C. 7290	20 54.4	45 58	6.03	.04	4.938	6
109	B.A.C. 7294	20 55.0	39 58	5.55	.06	4.941	6
110	59	20 56.1	42 56	4.74	.02	5.423	5...6
111	60	20 57.3	44 16	5.35	0.06	4.946	6
112	ξ	21 0.9	46 31	4.04	.01	5.423	4
113	L. 40951	21 1.8	59 16	6.07	.04	4.910	6
114	61	21 2.0	51 49	4.98	.04	2.381	5...6
115	63	21 2.8	42 48	5.30	.03	4.941	5...6
116	P. xxi. 1	21 3.9	60 16	5.49	0.09	4.910	6
117	B.A.C. 7365	21 6.9	36 53	5.96	.03	4.956	6
118	ζ	21 8.3	60 15	3.09	.c6	...	3
119	τ	21 10.4	52 25	3.65	.04	...	4
120	L. 41422	21 13.1	47 46	5.97	.03	4.912	6

No. 99. Σ 2726, Dist. 6″.6. Observed as one mass.
No. 100. Birm., No. 570. The colour is decidedly yellow.
No. 101. OΣ 413, Dist. <1″.0. Rapid binary. Observed as one mass. 4.90 is the mean of two determinations (4.93, 4.87) made with twenty extinctions on 1882.455, and with six on 1885.415. H.P. = 4.57.
No. 102. 5.39 is the mean of two determinations (5.40, 5.38) made on 1884.932 and 1885.415, each with six extinctions. H.P. = 5.00.
No. 110. Σ 2743, Dist. 20″. The brighter star observed.
No. 112. Birm., No. 576. The colour is slightly orange. The observation of 1882.430 (see Memoirs R.A.S., vol. xlvii, p. 432) is rejected. See note to No. 63.
No. 114. Σ 2758, Dist. 15″. Probably binary. Parallax 0.56. (Auwers.) Observed as one mass.
No. 116. Σ 2762, Dist. 3″.5. Observed as one mass.
No. 118. 3.09 is the mean of two determinations (3.04, 3.14) made on 1882.455 with twenty extinctions, and on 1885.423 with six. H.P. = 3.48.
No. 119. Double. Dist. 1″. Rapid binary. Observed as one mass. 3.65 is the mean of two determinations (3.68, 3.62) made on 1882.455 with twenty extinctions, and on 1885.423 with six.

Reference Number.	Star's Designation.	R.A. 1890.	N.P.D. 1890.	Adopted Zenithal Magnitude. Polaris 2.05.	Average Deviation in Magnitude.	Date, 1880 +	Mag. Argel. Uran.
		h. m.	o ,				
121	σ	21 13.1	51 4	4.52	0.06	2.591	4...5
122	υ	21 13.4	55 34	4.33	.07	2.591	4...5
123	68	21 14.3	46 31	5.04	.04	4.938	5
124	B.A.C. 7411	21 15.7	40 57	5.49	.07	4.956	5...6
125	L. 41554	21 16.6	57 51	6.10	.04	4.910	6
126	B.A.C. 7431	21 18.2	41 5	5 83	0·03	4.941	6 ..5
127	69	21 21.3	53 49	5.97	.09	4.910	6
128	B.A.C. 7455	21 21.3	43 46	5.68	.06	4.956	6
129	70	21 22.8	53 22	5.07	.05	4.912	6...5
130	O.A. 22275	21 22.9	41 38	5.31	.03	4.956	5
131	B.A.C. 7465	21 23.4	58 15	5.89	0.10	4.924	6
132	71 (g)	21 25.4	43 57	5.09	.10	4.956	5
133	B.A.C. 7483	21 26.7	37 32	6.11	.01	4.941	6
134	ρ....................	21 29.8	44 54	4.10	.07	2.591	4...5
135	72	21 30.3	51 58	5.25	.04	4.924	5
136	74	21 32.5	50 5	5.21	0.04	4.924	5
137	75	21 35.8	47 14	5.50	.13	4.956	6...5
138	π¹	21 38.2	39 19	4.92	.05	5.423	5...4
139	B.A.C. 7565	21 38.7	49 21	5.55	.07	4 956	5...6
140	79	21 38.9	52 13	5.68	.04	4.924	5 ..6
141	μ	21 39.2	61 45	4.50	0.08	2.591	4...5
142	π²	21 42.7	41 12	4.76	.11	5.423	4...5
143	L. 42563	21 43.8	51 51	5.94	.05	4.924	6
144	L. 42607	21 45.0	49 22	6.29	.03	4.941	6
145	B.A.C. 7631	21 48.4	34 43	5.58	.07	4 956	6
146	B.A.C. 7676	21 57.8	37 39	5.85	0.14	4.956	6

No. 138. The observation of 1882.591 (see Memoirs R.A.S., vol. xlvii, p. 432) is rejected on account of uncertainty introduced by ' passing clouds.'

No. 141. Σ 2822, Dist. 5".5. Observed as one mass.

No. 142. The observation of 1882.591 (see Memoirs R.A.S., vol. xlvii, p. 432) is rejected. See Note to No. 138.

No. 143. The star observed does not agree well with Argelander's place.

Nos. 145 and 146. The B.A.C. assigns these stars to Cepheus.

DELPHINUS.

Reference Number.	Star's Designation.	R.A. 1890.	N.P.D. 1890.	Adopted Zenithal Magnitude. - Pol r'e 2.05.	Average Deviation in Magnitude.	Date, 1880 +	Mag Argel. Uran.
		h. m.	° ′				
1	W.B. xx. 302......	20 14.3	77 6	6.38	0.03	3.609	6
2	L. 39188	20 17.8	75 49	5.94	.03	...	6
3	ε..............	20 28.0	79 4	3.59	.02	...	4
4	η............	20 28.7	77 21	5.32	.07	3.612	6...5
5	ζ............	20 30.2	75 42	4.80	.07	2.630	5...4
6	β	20 32.4	75 47	3.53	0.03	2.630	3...4
7	ι............	20 32.6	79 0	5.19	.06	3.609	6
8	θ	20 33.5	77 4	6.13	.07	3.615	6
9	κ	20 33.8	80 18	5.13	.05	3.615	5
10	α	20 34.5	74 29	3.93	.04	...	4...3
11	δ	20 38.3	75 19	4.59	0.09	2.630	4
12	γ¹	20 41.6	74 16	4.05	.07	2.630	} 3...4
13	γ²	20 41.6	74 16	4.09	.07	2.630	
14	13	20 42.4	84 24	5.72	.08	3.615	6
15	14	20 44.4	82 33	6.30	.03	3.609	6
16	15	20 44.4	77 52	6.15	0.06	3.645	6
17	16	20 50.4	77 51	5.28	.07	3.630	6...5
18	17	20 50.4	76 42	5.38	.10	3.645	6...5
19	18	20 53.1	79 35	5.56	.03	3.645	6...5
20	L. 40675	20 54.7	73 36	6.44	.03	3.609	6
21	L. 40682	20 55.4	71 5	5.02	0.03	3.645	6

No. 2. 5.94 is the mean of two determinations (5.93, 5.94) made on 1883.609 with twenty extinctions, and on 1885.442 with six. H.P. = 0.27.

No. 3. 3.59 is the mean of three accordant determinations, depending on twenty-six extinctions, observed 1882.630, 1885.442, and 1885.448. H.P. = 4.13.

Nos. 5, 7, 15, and 17. Each has the Spectrum I a ! (Vogel.)

No. 6. Σ 2704, Dist. 32″. The brighter star, which is really a close double, only observed. Spectrum I a ! (Vogel.)

No. 9. Called red in the Uranometria Argentina. The colour is not salient.

No. 10. Variable according to Auwers. (See Ast. Nach., vol. l, p. 106.) 3.93 is the mean of two identical determinations, each depending on twenty extinctions, made on 1882.630 and on 1882 638.

Nos. 12 and 13. These stars form Σ 2727, Dist. 12″.

No. 14. Variable according to Schmidt. (See Ast. Nach., vol. lxxiv, p. 286.)

No. 21. Spectrum III a !! (Vogel.)

DRACO.

Reference Number.	Star's Designation.	R.A. 1890.	N.P.D. 1890.	Adopted Zenithal Magnitude. Polaris 2.05.	Average Deviation in Magnitude.	Date, 1880+	Mag. Argel. Uran.
		h. m.	° ′				
1	P. ix. 37	9 21.3	8 11	4.43	0.05	2.708	4...5
2	Bradley 1446......	10 25.9	13 43	4.94	.09	2.708	5...4
3	B.A.C. 3747	10 51.2	11 39	6.25	.02	5.215	6
4	λ	11 24.9	20 4	3.80	.07	...	3...4
5	3....................	11 36.3	22 39	5.36	.11	5.283	5...6
6	R. 2794	11 59.7	12 29	5.82	0 12	5.294	6
7	B.A.C. 4112	12 7.1	11 46	5.12	.04	2.712	5...4
8	P. xii. 45	12 13.9	14 14	5.45	.18	5.215	6...5
9	4....................	12 25.3	20 11	5.18	.01	5.215	5
10	κ.................	12 28.8	19 36	3.68	.06	2.712	3...4
11	7....................	12 43.1	22 37	5.05	0.05	5.250	6
12	8	12 51.1	23 58	5.10	.04	5.223	5
13	9	12 55.8	22 48	5.63	.03	5.283	6
14	P. xii. 255	12 57.5	25 48	5.95	.05	5.223	6
15	R. 3020	13 23.4	24 41	6.09	.11	5.294	⎫ 6
16	R. 3021	13 23.5	24 42	6.52	0.09	5 294	⎭
17	P. xiii. 184........	13 38.1	24 37	5.58	.07	5.215	6
18	R. 3103	13 46.2	27 58	6.06	.02	5.250	6...5
19	10	13 48.2	24 44	4.82	.04	5.223	5
20	R. 3119	13 54.1	27 59	6.32	.01	5.247	6
21	a....................	14 1.4	25 6	3.56	0.08	2.712	3...4
22	D.M.+57°, No. 1498	14 12.4	32 48	6.49	.04	5.250	⎫ 6
23	D.M.+57°, No. 1499	14 12.6	32 48	6.63	.03	5.250	⎭
24	B.A.C. 4817	14 28.2	26 20	6.19	.13	5.247	6
25	R. 3225	14 28.7	29 18	6.13	.01	5.223	6

No. 4. 3.80 is the mean of two determinations (3.75, 3.84) made on 1882.712 with twenty extinctions, and on 1885.423 with six. Birm., No. 262. The star is of a deep yellow tint.

Nos. 9 and 13. Birm., Add. I, Nos. 44 and 46. These stars are decidedly orange.

No. 19. =87 Ursæ Majoris. See Introduction to B.A.C., p. 75. Birm., No. 315. The star is slightly red.

No. 23. Σ 1831, Dist. 6″. Observed as one mass.

Reference Number.	Star's Designation.	R.A. 1890.	N.P.D. 1890.	Adopted Zenithal Magnitude. Polaris 2.05.	Average Deviation in Magnitude.	Date, 1880 +	Mag. Argel. Uran.
		h. m.	° ′				
26	O.A. 14665	14 29.1	34 7	6.06	0.13	5.250	6
27	B.A.C. 4874	14 39.3	28 16	6.41	.04	5.247	6
28	P. xiv. 217.........	14 48.7	30 16	5.69	.07	5.247	6
29	B.A.C. 4967	14 58.9	29 22	5.85	.08	5.223	6
30	B.A.C. 4992	15 3.2	35 1	5.34	.03	5.239	6...5
31	B.A.C. 5071	15 16.9	37 41	5.87	0.12	5.247	6
32	B.A.C. 5091	15 20.8	26 16	5.88	.12	5.247	6
33	ι	15 22.5	30 39	3.26	.06	...	3
34	B.A.C. 5115	15 25.7	28 57	6.06	.05	5.239	6
35	P. xv. 110	15 25.7	27 21	6.05	.05	5.239	6
36	P. xv. 136	15 29.4	25 25	5.85	0.00	5.223	6
37	R. 3426	15 34.7	35 9	6.03	.01	5.250	6
38	B.A.C. 5181	15 35.4	39 13	5.85	.05	5.247	6
39	B.A.C. 5210	15 39.9	37 18	5.49	.09	5.239	6
40	B.A.C. 5248	15 45.0	34 17	5.90	.12	5.247	6...5
41	P. xv. 198	15 45.0	27 4	5.40	0.01	5.239	.5
42	B.A.C. 5279	15 49.7	33 51	6.12	.08	5.250	6
43	B.A.C. 5313	15 55.2	34 57	5.23	.05	...	5...6
44	θ	15 59.9	31 9	3.86	.05	...	4...3
45	B.A.C. 5406	16 6.0	21 54	5.50	.02	5.283	6...5
46	B.A.C. 5415	16 7.3	31 53	6.35	0.13	5.292	6
47	R. 3527	16 11.9	22 34	6.31	.05	5.292	6
48	B.A.C. 5459	16 15.4	29 58	5.57	.12	5.247	6
49	B.A.C. 5502	16 22.0	34 33	5.60	.04	5.261	6...5
50	B.A.C. 5514	16 22.1	20 38	5.26	.11	5.261	6...5

No. 27. Σ 1878, Dist. 3″. Observed as one mass.

Nos. 31 and 38. B.A.C. assigns these stars to Boötes.

No. 33. Birm., No. 352. The red colour is not salient. 3.26 is the mean of seven accordant determinations, made on as many nights, four at Oxford and three at Cairo, involving one hundred and ten extinctions, between 1882.712 and 1883.131.

No. 43. 5.23 is the mean of two determinations (5.30, 5.16) made on 1885.247 and 1885.423, each with six extinctions. H.P. = 5.02.

No. 44. 3.86 is the mean of two determinations (3.85, 3.86) made on 1882.712 with twenty extinctions, and on 1885.423 with six.

Reference Number.	Star's Designation.	R.A. 1890.	N.P.D. 1890.	Adopted Zenithal Magnitude. Polaris 2.05.	Average Deviation in Magnitude.	Date, 1880+	Mag. Arcel. Uran.
		h. m.	° ′				
51	η	16 22.5	28 14	2.79	0.07	2.717	3...2
52	R. 3566	16 25.9	38 21	6.36	.01	5.292	6
53	15	16 28.2	21 0	4.92	.09	5.292	5
54	P. xvi. 140........	16 30.9	28 57	5.71	.10	5.261	6
55	16	16 33.6	36 53	5.68	.07	5.423	}
							4...5
56	17	16 33.6	36 51	5.22	0.10	2.717	
57	B.A.C. 5599	16 35.8	33 46	5.53	.13	5.261	6...5
58	18	16 40.2	25 12	5.03	.05	5.294	5...6
59	B.A.C. 5643	16 43.2	33 1	4.90	.16	5.283	5
60	P. xvi. 219........	16 44.6	34 24	6.01	.08	5.294	6
61	19	16 55 4	24 42	4.92	0.19	5.292	5
62	μ	17 3 0	35 23	5.00	.07	2.717	5...4
63	L. 31378	17 8.1	37 28	6.16	.08	5.261	6
64	ζ................	17 8.5	24 9	3.29	.07	2.717	3
65	P. xvii. 61	17 11.6	27 0	5.45	.03	5.283	6
66	R. 3696	17 15.2	29 13	6.03	0.07	...	6
67	B.A.C. 5917	17 24.3	29 52	5.47	.06	...	6
68	β	17 28.0	37 37	2.06	.13	2.717	3...2
69	ν¹	17 30.0	34 44	4.75	.09	2.717	} 4
70	ν²	17 30.1	34 45	4.76	.03	2.717	
71	27	17 32.4	21 48	5.06	0.07	5.292	5...6
72	26	17 33.8	28 2	5.51	.02	5.283	6...5
73	ω	17 37.6	21 12	4.81	.10	5.283	5
74	L. 32455	17 38.8	38 7	6.13	.10	5.261	6
75	L. 32566	17 41.8	36 10	5.51	.05	...	6...5

No. 51. OΣ 312, Dist. 4″.7. Observed as one mass.

No. 55. This star is one of the components of Σ 30¹, Dist. 90″. The observation 1882.717 (see Mem. R.A.S., vol. xlvii, p. 433) has been rejected, owing to the observation having been made on an indifferent night.

No. 56. The other component of Σ 30¹, and is Σ 2078, Dist. 3″.7. Observed as one mass.

No. 58. Birm., Add. I, No. 49. The red colour is not salient.

No. 60. Identification doubtful. The place agrees with that given by Argelander.

No. 62. Σ 2130, Dist. 3″.5. Possibly binary. Observed as one mass.

No. 66. Σ 2155, Dist. 10″. The brighter star observed. 6.03 is the mean of two determinations (6.02, 6.04) made on 1885.294 and 1885.437, each with six extinctions.

No. 67. 5.47 is the mean of two determinations (5.50, 5.44) made on 1885.294 and 1885.437, each with six extinctions.

No. 68. Birm., No. 412. The red colour is not salient.

Nos. 69 and 70. Σ 35¹, Dist. 61″.

No. 75. 5.51 is the mean of two determinations (5.56, 5.46) made on 1885.261 and 1885.427, each with six extinctions.

Reference Number.	Star's Designation.	R.A. 1890.	N.P.D. 1890.	Adopted Zenithal Magnitude. Polaris 2.05.	Average Deviation in Magnitude.	Date, 1880 +	Mag. Arzel. Uran.
		h. m.	° ′				
76	ψ^1	17 43.9	17 48	4.84	0.06	2.731	} 4...5
77	ψ^2	17 43.9	17 48	5.80	.08	2.731	
78	30	17 46.5	39 12	5.17	.06	5.294	5
79	ξ	17 51.6	33 7	3.90	.06	2.723	3...4
80	γ	17 54.1	38 30	2.40	.04	2.723	2...3
81	35	17 54.4	13 1	5.08	0.13	5.292	5
82	34	17 57.1	17 59	5.57	.09	5.303	6
83	B.A.C. 6185	18 8.2	35 45	5.82	.06	5.303	6
84	40	18 8.3	10 1	5.69	.01	5.292	} 5
85	41	18 8.4	10 1	6.04	.04	5.292	
86	36	18 13.2	25 38	4.98	0.12	5.303	5
87	37	18 15.9	21 17	5.97	.13	5.253	6
88	B.A.C. 6246	18 17.4	38 42	6.15	.11	5.294	6
89	B.A.C. 6255	18 18.7	40 56	5.13	.07	5.305	5
90	39	18 22.3	31 16	4.84	.14	5.292	5
91	ϕ	18 22.3	18 43	4.22	0.04	2.731	4...5
92	χ	18 23.0	17 19	3.93	.04	2.731	4...3
93	42	18 25.7	24 30	5.08	.00	5.253	5
94	45	18 30.7	33 2	4.92	.13	5.294	5
95	B.A.C. 6350	18 31.4	37 44	5.33	.02	5.253	5...6
96	B.A.C. 6375	18 35.0	12 31	5.51	0.05	5.261	6
97	L. 34817	18 35.8	24 36	5.90	.05	5.261	6
98	O.A. 18518	18 36.5	27 34	5.58	.01	5.303	6
99	P. xviii. 170	18 37.4	37 54	5.65	.05	5.253	6
100	B.A.C. 6393	18 40.0	27 22	6.02	.06	5.305	6
101	46	18 40.5	34 34	5.31	0.07	5.305	5...6
102	B.A.C. 6428	18 45.4	41 22	5.07	.10	5.261	6
103	B.A.C. 6469	18 48.5	16 2	5.60	.00	5.305	6
104	B.A.C. 6452	18 49.1	37 10	5.44	.06	5.253	6
105	ϲ	18 49.6	30 45	4.72	.09	2.731	5...4

Nos. 76 and 77. Σ 2241, Dist. 31″.
No. 79. Birm., No. 421. The colour is slightly orange.
No. 80. Birm., No. 423. The red colour is not salient.
Nos. 84 and 85. Σ 2308, Dist. 20″.
No. 90. Σ 2323, Triple. The close pair, Dist. 3″. Observed as one mass.
No. 91. OΣ 353, Dist. <1″. Observed as one mass.
No. 95. Σ 2348, Dist. 26″. The brighter star observed.
No. 105. Σ 2420, Dist. 30″. The brighter star observed.

Reference Number.	Star's Designation.	R.A. 1890.	N.P.D. 1890.	Adopted Zenithal Magnitude. Polaris 2.05.	Average Deviation in Magnitude.	Date, 1880+	Mag. Argel. Uran.
		h. m.	o '				
106	50	18 49.9	14 42	5.60	0.05	3.253	6
107	B.A.C. 6470	18 50.5	39 26	5.12	.01	5.303	5
108	P. xviii. 254	18 51.9	41 17	5.77	.03	5.294	6
109	48	18 54.9	32 26	5.58	.05	5.253	6
110	ν	18 55.8	18 51	5.06	.05	5.305	5...6
111	O.A. 18836	18 56.1	24 52	5.83	0.09	5.303	6
112	L. 35681	18 57.4	39 37	4.99	.06	...	5...6
113	49	18 58.6	34 28	5.67	.15	5.253	6
114	51	19 2.5	36 46	5.40	.07	5.253	6...5
115	53	19 9.6	33 20	5.26	.02	5.294	6...5
116	54	19 12.0	32 29	5.11	0.04	5.303	5...6
117	δ	19 12.5	23 32	2.96	.07	2.731	3
118	59	19 13.2	13 37	5.17	.13	5.305	6..5
119	τ	19 17.7	16 51	4.64	.06	2.731	5
120	B.A.C. 6640	19 18.3	32 34	5.85	.07	5.294	6
121	D.M.+64°, No. 1344	19 19 1	25 49	6.20	0.11	5.303	6
122	R. 4304	19 19.5	32 28	6.50	.00	5.294	6
123	π	19 20.1	24 30	4.84	.04	...	5
124	σ	19 32.5	20 31	4.78	.05	5.294	5...6
125	B.A.C. 6808	19 44.5	20 56	5.85	.10	5.305	6
126	ε..................	19 48.5	20 1	3.72	0.08	2.731	4
127	64	20 0.3	25 29	5.50	.06	5.305	6...5
128	ρ	20 2.3	22 26	4.54	.08	5.294	5
129	66	20 3.8	28 20	5.54	.11	5.305	6
130	68	20 9.8	28 15	5.72	.06	5.303	6
131	B.A.C. 7017	20 16.3	23 31	5.98	0.13	5.294	6
132	71	20 17.8	28 6	5.77	.03	5.305	6
133	B.A.C. 7037	20 19.6	21 28	5.92	.02	5.305	6
134	73	20 33.0	15 25	5.41	.05	...	5...6
135	75	20 35.2	8 57	5.78	.11	5.305	6

No. 112. 4.99 is the mean of two determinations (5.06, 4.93) made on 1885.294 and on 1885.437, each with six extinctions. H.P.=5.37.

No. 117. Birm., No. 491. The star is yellow.

No. 120. The star observed is the preceding of two, both near Argelander's place.

No. 123. 4.84 is the mean of two determinations (4.79, 4.88) made on 1885.253 and on 1885.437, each with six extinctions. H.P.=4.55.

No. 126. Σ 2603, Dist. 3″. Observed as one mass. Probably binary.

No. 134. 5.41 is the mean of two determinations (5.46, 5.36, made on 1885.294 and on 1885.437, each with six extinctions. H.P.=5.20.

Reference Number.	Star's Designation.	R.A. 1890.	N.P.D. 1890.	Adopted Zenithal Magnitude. Polaris 2.05.	Average Deviation in Magnitude.	Date, 1880 +	Mag. Argel. Uran.
		h. m.	o ,				
136	76	20 50.6	7 52	5.71	0.03	5.294	6
137	B.A.C. 7299	20 52.6	9 51	5.20	.13	5.305	6

No. 137. The B.A.C. assigns this star to Cepheus.

EQUULEUS.

1	P. xx. 376	20 50.2	85 53	6.15	0.03	...	6
2	1.....................	20 53.6	86 8	5.41	.06	3.628	5
3	2.....................	20 56.8	83 15	6.47	.03	3.628	6
4	3.....................	20 59.1	84 56	5.70	.05	3.628	6
5	D.M.+1°, No. 4418	20 59.2	88 9	6.41	.01	3.645	6
6	4.....................	21 0.0	84 29	5.76	0.02	...	6
7	γ.....................	21 5.0	80 19	4.39	.08	...	5...4
8	δ	21 9.1	80 26	4.51	.04	2.630	5...4
9	a.....................	21 10.3	85 15	3.92	.03	2.630	4
10	9.....................	21 15.6	83 7	5.84	.03	3.628	6
11	D.M.+2°, No. 4348	21 17.0	87 32	6.23	0.04	3.645	6
12	β	21 17.4	83 40	5.00	.11	3.639	5
13	D.M. +9°, No. 4800	21 18.9	80 18	6.33	.06	3.639	6

No. 1. Σ 2735, Dist. 2″. Observed as one mass. 6.15 is the mean of two determinations (6.28, 6.03) made on 1883.628 with twenty extinctions, and on 1885.504 with six. H.P. = 5.90.

No. 2. Triple. The close pair, Σ 2737, Dist. > 1″, observed as one mass.

No. 3. Σ 2742, Dist. 2″. Observed as one mass. The star observed agrees with Argelander's place, but Gould suggests that a brighter star preceding two minutes in R.A. was the object seen by Argelander.

No. 4. Called red in the Uranometria Argentina.

No. 6. 5.76 is the mean of two determinations (5.71, 5.81) made on 1883.628 with twenty extinctions, and on 1885.504 with six. H.P. = 6.18.

No. 7. The star observed is the brighter component of Σ 54¹. There are several minute companions. 4.39 is the mean of two determinations (4.34, 4.44) made on 1883.628 with twenty extinctions, and on 1885.504 with six. H.P. = 4.79. Spectrum I a! (Vogel.)

No. 8. Σ 2777 and OΣ 535 ; the latter of which is observed as one mass : Dist. > 1″. Binary : the period of possibly 11 years is the shortest yet computed.

No. 9. Spectrum I a! (Vogel.)

No. 10. Spectrum III a! (Vogel.)

No. 11. Several stars of approximately equal magnitude near.

No. 12. Quadruple. The brightest star only observed.

ERIDANUS.

Refer-ence Number.	Star's Designation.	R.A. 1890.	N.P.D. 1890.	Adopted Zenithal Magnitude. Polaris =2.05.	Average Deviation in Magnitude.	Date, 1880 +	Mag. Argel. Uran.
		h. m.	° ′				
1	η...................	2 51.0	99 20	3.87	0.07	5 022	3
2	L. 5449	2 51.1	94 9	5.23	.01	5.025	6...5
3	L. 5516	2 53.2	93 13	5.40	.01	5.061	6
4	5...................	2 54.1	92 54	5.32	.02	5.025	5...6
5	ρ²	2 57.3	98 7	5.52	.10	5.022	6
6	10	2 58.9	98 2	5.69	0.11	5.018	6
7	W.B. ii. 1054 ...	3 1.1	96 31	5.74	.07	5.025	6
8	ζ...................	3 10.5	99 14	4.82	.04	5.022	4...5
9	W.B. iii. 147	3 10.6	96 20	6.35	.10	5.061	6
10	14	3 11.3	99 34	6.10	.08	5.018	6
11	L. 6462	3 24.2	97 11	6.22	0.04	5.061	6
12	17	3 26.2	95 27	4.72	.07	5.025	5
13	ε	3 27.5	99 51	3.45	.02	5.022	3
14	L. 6726	3 33.1	97 45	6.22	.10	5.042	6
15	W.B. iii. 604	3 34.1	93 45	6.47	.07	5 061	6
16	22	3 35.2	95 34	5.41	0.09	5.042	6
17	24	3 38.9	91 30	5.40	.08	5.018	6
18	30	3 47.3	95 41	5.39	.07	5.025	6...5
19 {	32 1st star	3 48.8	93 17	6.44	.01	5.025	} 5
	32 2nd star........	3 48.8	93 17	4.83	.09	5 025	
20	L. 7384	3 53.5	95 47	6.10	.09	5.018	6
21	35	3 55.9	91 51	5.10	0.05	5.022	5...6
22	L. 7484	3 57.0	90 34	5.60	.04	5.025	6
23	o¹	4 6.5	97 7	4.10	.06	5.018	4...5
24	o²	4 10.3	97 47	4.54	.05	5.042	5...4
25	L. 8048	4 11.9	96 44	6.25	.07	5.061	6

No. 2. Variable according to Schmidt. (See Ast. Nach., xcv. p. 366.)
No. 5. Double. Dist. 2″.5. Observed as one mass.
Nos. 16 and 22. D'Arrest suggests variability. (See Notes to Berlin Charts.)
Nos. 17 and 21. Spectrum I a! (Vogel.) D'Arrest suggests variability.
No. 18. A faint companion not observed.
No. 19. Σ 470, Dist. 6″.
No. 24. Σ 518, Dist. 84″. The faint companion not observed. Rapid common proper motion.
The large star is itself double, Dist. 3″. Observed as one mass. Binary.

Reference Number.	Star's Designation.	R.A. 1890.	N.P.D. 1890.	Adopted Zenithal Magnitude. Polaris 2.05.	Average Deviation in Magnitude.	Date, 1880 +	Mag. Argel. Uran.
		h. m.	° ′				
26	L. 8154	4 15.2	96 30	6.27	0.08	5.061	6
27	d	4 15.4	97 51	5.77	.03	5.018	6
28	ξ	4 18.2	94 0	5.36	.06	5.025	5...6
29	44	4 29.8	88 52	5.64	.05	5.042	6
30	45	4 26.2	90 17	5.07	.08	5.061	5...6
31	46	4 28.6	96 58	5.42	0.12	5.061	6...5
32	47	4 28 9	98 28	5.24	.01	5.018	6...5
33	W.B. iv. 585	4 28 9	99 11	5.96	.19	5.061	6
34	ν	4 30.9	93 34	3.07	.05	5.025	3...4
35	49	4 31.5	89 13	5.40	.03	5.018	6...5
36	51	4 32.1	92 41	5.54	0.01	5.042	5...6
37	55	4 38.3	99 0	5.96	.06	5.061	6
38	56	4 38.8	98 42	5.76	.04	5.042	6
39	μ	4 40.0	93 27	4.12	.02	5.018	4...3
40	ω	4 47.5	95 38	4.27	.07	5.025	4...5
41	62	4 51.0	95 21	5.66	0.03	5.061	6
42	ψ	4 56 1	97 20	4.79	.00	5.061	5...4
43	β	5 2.4	95 14	2.85	.07	5.042	3
44	68	5 3.3	94 36	5.34	.09	5.061	6
45	λ	5 3.9	98 54	4.29	.01	5.018	4

No. 26. Birm., No. 73. The red colour is not salient.
Nos. 30 and 33. Called red in Uranometria Argentina.
No. 32. Birm., Add. I, No. 14. The colour of the star is slightly red.
No. 35. Spectrum I a! (Vogel.)
No. 36. Probably variable. (See Uranometria Argentina, p. 273.)
No. 37. Σ 590, Dist. 10″. Observed as one mass.
No. 41. Has a faint distant companion, 63″, not observed.
No. 43. Probably variable. (See Uranometria Argentina, p. 273.)
No. 45. Gilliss suspects this star of variability. (See Ast. Obs., p. 663.)

GEMINI.

Reference Number.	Star's Designation.	R.A. 1890.	N.P.D. 1890.	Adopted Zenithal Magnitude. Polaris 2.05.	Average Deviation in Magnitude.	Date, 1880 +	Mag. Argel. Uran.
		h. m.	° ′	″			
1	1	5 57.4	66 44	4.53	0.06	4.272	5
2	η	6 8.2	67 28	3.65	.03	2.300	3...4
3	μ	6 16.3	67 26	3.45	.09	2.300	3
4	ν	6 22.4	69 43	4.33	.04	...	5 4
5	γ	6 31.4	73 30	2.13	.05	...	2...3
6	26	6 36.0	72 15	5.20	0.06	4.272	6...5
7	ε	6 37.2	64 46	3.29	.04	2.300	3 4
8	28	6 37 8	60 55	5.38	.08	4.242	6
9	30	6 37.8	76 40	5.19	.03	...	5
10	ξ²	6 39.1	76 59	3.84	.04	2.305	4 3
11	W.B. vi. 1227 ...	6 42.6	57 16	5.87	0.06	4.242	6
12	33	6 43.5	73 40	5.75	.07	4.275	6
13	36	6 45.0	68 6	5.30	.04	4.275	6
14	θ	6 45.5	55 54	3.60	.05	2.305	3...4
15	38	6 48.4	76 41	4.86	.02	4.242	5
16	41	6 54 0	73 46	6.12	0.04	4.275	6
17	ω	6 55.7	65 38	5.30	.06	4.312	6
18	B.A.C. 2306	6 57.5	78 53	5.22	.03	4.275	6
19	ζ	6 57.6	69 16	4.01	.06	2.305	4
20	45	7 2.1	73 53	5.67	.03	4.242	6

No. 2. Variability discovered by Schmidt in 1865. Varies from 3.2 to 4.2 mag. Period 229.1 days. Epoch of minimum 1870, Apr. 7.

No. 3. Birm., No. 143. The star is slightly red.

No. 4. 4.33 is the mean of two determinations (4.39, 4.26) made on 1882.300 with twenty extinctions, and on 1885.239 with six. H.P.=3.98.

No. 5. 2.13 is the mean of two determinations (2.14, 2.13) made on 1882.300 with twenty extinctions, and on 1883.193 with ten. Spectrum I a !! (Vogel.)

No. 6. Spectrum I a !! (Vogel.)

No. 7. Birm., No. 152. The red colour is not salient.

No. 9. 5.19 is the mean of two determinations (5.22, 5.16) made on 1884.275 with twenty extinctions, and on 1885.253 with six. H.P.=4.62. Spectrum II a ! (Vogel.)

No. 10. The determination of 1882.305 (see Memoirs R.A.S., vol. xlvii, p. 435) is rejected. The sky was covered with haze at the time. Spectrum I a ! (Vogel.)

No. 12. This magnitude was exactly reproduced 1885.239. H.P.=5.44.

No. 15. Σ 982, Dist. 6″. Binary. Observed as one mass. Struve thought the variability certain. (See Mens. Mic., p. lxxiii.)

No. 18. The B.A.C. assigns this star to Monoceros. Spectrum II a ! (Vogel)

No. 19. Variability discovered by Schmidt in 1847. Varies from 3.7 to 4.5 mag. Period 10ᵈ, 3ʰ 43ᵐ 12ˢ. Epoch of maximum 1863, July 17. 4ʰ 52ᵐ. Epoch of minimum 1863, July 12, 4ʰ 30ᵐ.

Reference Number.	Star's Designation.	R.A. 1890.	N.P.D. 1890.	Adopted Zenithal Magnitude. Polaris 2.05.	Average Deviation in Magnitude.	Date, 1880+	Mag. Argel. Uran.
		h. m.	° ′				
21	τ...............	7 4.1	59 34	4.71	0.09	2.305	5...4
22	47	7 4.6	62 58	5.63	.07	4.242	6
23	51	7 7.1	73 39	5.62	.11	4.242	6
24	L. 14100	7 11.1	58 51	6.06	.05	4.275	6
25	λ	7 11.8	73 16	3.72	.05	2.305	4...3
26	δ...............	7 13.6	67 49	3.48	0.09	2.305	3...4
27	56	7 15.5	69 21	5.14	.02	4.275	6...5
28	57	7 16.8	64 44	5.20	.04	4.275	5...6
29	58	7 16.9	66 51	6.09	.03	4.281	6
30	ι...............	7 18.9	61 59	3.98	.06	2.305	4
31	61	7 20.5	69 31	5.51	0.06	4.281	6
32	63	7 21.2	68 20	5.23	.03	4.275	6...5
33	ρ...............	7 22.0	58 0	4.45	.04	...	5
34	64	7 22.5	61 39	5.13	.03	5.239	5
35	65	7 23.0	61 51	4.97	.04	4.281	5
36	W.B. vii. 685......	7 25.4	72 41	5.81	0.04	4 297	6
37	68	7 27.3	73 56	5.21	.05	4.297	6...5
38	α...............	7 27.6	57 52	1.53	.05	...	2...1
39	v...............	7 29.2	62 52	4.37 .	.09	...	4...5
40	70	7 31.3	54 42	5.82	.04	4.275	6
41	ο...............	7 32.0	55 10	5.02	0.09	...	5...6
42	74	7 33.1	72 4	5.20	.09	4.275	6
43	σ	7 36.4	60 51	4.06	.05	4.305	5
44	76	7 37.4	63 57	5.34	.04	4.297	6
45	κ...............	7 37.8	65 20	3.63	.07	2.311	4...3

No. 23. Spectrum III a !!! (Vogel.)

No. 25. Σ 1061, Dist. 10″. Observed as one mass. Probably variable. (See Ast. Nach., xciv. p. 243.)

No. 26. Σ 1066, Dist. 6″. Observed as one mass.

No. 33. 4.45 is the mean of two determinations (4.50, 4.40) made on 1882.305 with twenty extinctions, and on 1885.239 with six. H.P. = 4.15.

No. 38. Σ 1110, Dist. 6″. Binary. Observed as one mass. 1.53 is the mean of seven accordant determinations made on as many nights; four at Oxford and three at Cairo, involving eighty extinctions, between 1882.979 and 1883.205. H.P. = 1.56.

No. 39. Birm., No. 181. The star is orange. 4.37 is the mean of two identical determinations, each depending on twenty extinctions, made 1882.311 and 1882.316.

No. 41. 5.02 is the mean of two determinations (5.01, 5.03) made on 1884.297 with twenty extinctions, and on 1885.239 with six. H.P. = 4.67.

No. 42. Spectrum III a ! (Vogel.)

No. 43. Birm., No. 183. The red colour is not salient.

Reference Number.	Star's Designation.	R.A. 1890.	N.P.D. 1890.	Adopted Zenithal Magnitude. Polaris 2.05.	Average Deviation in Magnitude.	Date, 1880+	Mag. Argel. Uran.
		h. m.	° ′				
46	β	7 38.6	61 42	1.36	0.07	...	1...2
47	81	7 39.8	71 13	5.34	.05	4.297	6...5
48	π	7 40.4	56 19	5.55	.09	4.275	6
49	φ	7 46.8	62 57	4.97	.03	4.297	5
50	85	7 49.2	69 49	5.19	.05	4.281	6...5
51	χ	7 56.8	61 54	5.09	0.04	4.297	5
52	ψ	8 6.3	60 1	5.65	.08	4.297	6

No. 46. Σ 5ʰ. The very faint distant companion not observed. Birm., No. 188. The red colour is not salient. 1.36 is the mean of seven accordant determinations, made on as many nights; four at Oxford and three at Cairo, involving eighty extinctions, between 1882.979, and 1883.206. H.P.=1.12.
No. 48. Σ 1135, Dist. 22″. The brighter star observed.
No. 51. Called 6 Cancri in B.A.C. and Nautical Almanac.
No. 52. Called 15 Cancri in B.A.C.

HERCULES.

1	P. xv. 153	15 34.7	42 50	5.93	0.02	4.543	6
2	χ	15 48.8	47 15	4.48	.07	2.376	4...5
3	2...............	15 51.0	46 35	5.63	.04	4.541	6
4	4...............	15 51.8	47 7	5.92	.05	4.553	6
5	5 (r)	15 56.3	71 53	5.24	.08	4.541	6...5
6	υ...............	15 59.4	43 39	4.00	0.03	2.376	4...5
7	W.B. xv. 1569 ...	16 2.6	67 53	6.48	.04	4.560	6
8	κ...............	16 3.1	72 40	5.04	.05	4.541	5
9	φ	16 5.3	44 47	4.10	.10	2.376	4
10	P. xvi. 12	16 6.5	73 3	6.05	.04	4.563	6
11	10	16 7.0	66 13	6.00	0.08	4.553	6
12	9...............	16 7.8	84 42	5.88	.04	4.553	6
13	B.A.C. 5452	16 15.3	68 36	6.28	.05	4.553	6
14	B.A.C. 5460	16 16.2	50 2	5.71	.07	4.560	6
15	τ...............	16 16.4	43 26	3.63	.13		3...4

No. 1. B.A.C. assigns this star to Boötes.
No. 7. Birm., No. 369. The red colour is not salient.
No. 8. Σ 2010, Dist. 31″. The brighter star observed.
No. 15. 3.63 is the mean of two determinations (3.55, 3.71) made on 1882.305 with twenty extinctions, and on 1885.333 with six.

Reference Number.	Star's Designation.	R.A. 1890.	N.P.D. 1890.	Adopted Zenithal Magnitude. Polaris 2.05.	Average Deviation in Magnitude.	Date, 1880 +	Mag. Argel. Uran.
		h. m.	° ′				
16	γ	16 17.1	70 35	3.56	0.08	2.387	3
17	21	16 18.7	82 48	5.94	.04	4.560	6
18	23	16 18.7	57 25	6.15	.04	4.553	6
19	ω	16 20.3	75 43	4.70	.08	4.541	5
20	25	16 21.5	52 21	5.63	.06	4.543	6...5
21	30 (*g*).............	16 25.0	47 53	5.14	0.09	4.560	5...6
22	β	16 25.5	68 16	2.67	.06	2.387	2...3
23	B.A.C. 5527	16 25.8	69 17	4.91	.04	4.553	6
24	28 (*u*)	16 27.2	84 15	5.70	.08	4.560	6
25	29 (*h*)	16 27.5	78 16	5.18	.03	4.541	5...6
26	W.B. xvi. 840 ...	16 28.7	44 5	5.52	0.08	4.563	6
27	σ	16 30.6	47 20	4.06	.05	2.376	4
28	B.A.C. 5568	16 33.0	43 10	6.07	.06	4.560	6
29	36	16 35.1	85 35	6.09	.09	4.560	} 6
30	37	16 35.2	85 34	5.80	.07	4.560	
31	42	16 35.8	40 51	4.95	0.06	4.543	5...4
32	ζ.................	16 37.2	58 12	2.04	.06	2.376	3...2
33	η	16 39.1	50 52	3.60	.08	2.376	3
34	P. xvi. 177........	16 39.8	55 46	5.91	.08	4.566	6
35	B.A.C. 5620	16 40.4	74 3	5.75	.06	4.563	6

No. 16. The faint distant companion not observed. Gilliss suggests variability. (See Ast. Obs., p. 667.) Spectrum I a! (Vogel.) A second determination on 1885.333 with six extinctions, gave the magnitude 3.66. H.P. = 3.83.

No. 17. Spectrum I a! (Vogel.)

No. 19. Called 51 Serpentis. (See Introd. to B.A.C., p. 75.) Spectrum I a! (Vogel).

No. 21. Birm., No. 382. The star is yellow or slightly orange. Baxendell detected variability in 1857 from 5.0 to 6.2 mag. Period irregular.

No. 22. Birm., No. 383. The red colour is not salient.

No. 24. Called 11 Ophiuchi. (See Introd. to B.A.C., p. 75.)

No. 26. Σ 2063, Dist. 16″. Observed as one mass.

Nos. 29 and 30. These stars form Σ 31¹, Dist. 69″.

No. 30. Spectrum I a! (Vogel.)

No. 31. Σ 2082, Dist. 22″. The brighter star observed.

No. 32. Σ 2084, Binary. Period 35 years. Observed as one mass.

No. 35. Spectrum III a! (Vogel.)

Reference Number.	Star's Designation.	R.A. 1890.	N.P.D. 1890.	Adopted Zenithal Magnitude. Polaris 2.05.	Average Deviation in Magnitude.	Date, 1880 +	Mag. Argel. Uran.
		h. m.	o '				
36	43	16 40.6	81 13	5.74	0.08	4.541	6...5
37	45	16 42.3	84 33	5.43	.03	4.541	6
38	B.A.C. 5647	16 44.4	76 32	6.13	.07	4.563	6
39	47	16 45.0	82 34	5.49	.01	4.541	6...5
40	52	16 46.4	43 49	5.02	.02	4.549	4...5
41	50	16 46.4	60 1	6.04	0.05	4.563	6
42	49	16 47.1	74 51	6.51	.03	4.566	6
43	51	16 47.2	65 10	5.02	.01	4.543	6
44	53	16 48.8	58 7	5.43	.05	4.563	5
45	W.B. xvi. 1513 ...	16 50.1	68 50	5.58	.09	4.541	6
46	54	16 50.5	71 23	5.29	0.07	4.541	6
47	ε....................	16 56.1	58 55	3.85	.09	...	3...4
48	W.B. xvi. 1688 ...	16 56.3	67 12	5.92	.06	4.563	6
49	59	16 57.6	56 16	5.21	.08	4.543	5
50	W.B. xvi. 1735 ...	16 57.8	64 20	6.21	.06	4.563	6
51	P. xvi. 279.........	16 58.1	75 45	5.35	0.03	4.560	5
52	P. xvi. 283.........	16 58.6	76 14	6.20	.07	4.566	} 6
53	P. xvi. 285.........	16 58.9	76 16	6.28	.09	4.566	
54	P. xvi. 292.........	16 59.6	70 15	6.08	.02	4.566	6
55	W.B. xvi. 1807 ...	16 59.9	55 3	5.98	.07	4.566	6
56	60	17 0.3	77 6	4.95	0.03	4.543	5
57	W.B. xvi. 1844 ...	17 1.5	67 46	6.09	.05	4.579	6
58	P. xvii. 3	17 4.1	53 55	5.58	.03	4.549	5
59	B.A.C. 5790	17 4.2	49 20	6.09	.05	4.563	6
60	W.B. xvii. 114 ...	17 6.0	49 5	5.33	.10	4.549	5

No. 36. Birm., No. 387. The star is slightly orange. Double. The distant faint companion not observed. Spectrum II a !! (Vogel.)

No. 38. Σ 2103, Dist. 5".5. Observed as one mass. Spectrum I a ! (Vogel.)

No. 39. Spectrum I a ! (Vogel.)

No. 46. Spectrum II a ! (Vogel.)

No. 47. 3.85 is the mean of three accordant determinations involving forty extinctions made on 1882.376, 1882.387, and 1882.738.

No. 51. Spectrum III a ! (Vogel.)

Nos. 52 and 53. These stars form Σ 33¹, Dist. 290".

Reference Number.	Star's Designation.	R.A. 1890.	N.P.D. 1890.	Adopted Zenithal Magnitude. Polaris 2.05.	Average Deviation in Magnitude.	Date, 1880 +	Mag. Argel. Uran.
		h. m.	° ′				
61	63	17 6.5	65 38	6.20	0.08	4.579	6
62	a	17 9.6	75 25	3.02	.04	2.387	Var.
63	δ....................	17 10.5	65 2	3.25	.02	5.333	3
64	π....................	17 11.2	53 4	3.60	.05	2.376	3...4
65	68	17 13.3	56 47	5.14	.11	4.543	5
66	69	17 13.9	52 36	4.52	0.06	2.376	5
67	P. xvii. 69	17 14.0	40 12	6.14	.08	4.563	6
68	P. xvii. 64	17 14.4	61 5	6.01	.06	4.549	6
69	W.B. xvii. 377 ...	17 14.7	51 4	6.07	.07	4.579	6
70	P. xvii. 68	17 15.5	71 50	4.99	.08	4.579	6
71	P. xvii. 71	17 15.6	64 21	5.51	0.09	4.563	6
72	70	17 16.4	65 23	5.56	.04	4.543	6
73	72	17 16.5	57 23	5.51	.05	4.546	5...6
74	74	17 17.2	43 39	5.66	.04	4.549	6...5
75	B.A.C. 5874	17 18.1	49 55	5.40	.06	...	5
76	73	17 19.5	66 56	5.94	0.05	4.566	6
77	ρ....................	17 19.9	52 45	4.35	.07	2.376	4
78	W.B. xvii. 554 ...	17 20.3	51 17	6.47	.03	4.582	6
79	P. xvii. 109	17 22.1	69 50	5.62	.07	4.579	6
80	W.B. xvii. 663 ...	17 22.7	55 11	6.09	.09	4.582	6

No. 62. Σ 2140, Dist. 4″.5. Observed as one mass. Birm., No. 402. The colour is orange. Variable from 3.0 to 3.9 mag. Period irregular. The mean of two accordant determinations at Cairo, 1883.138, and 1883.160, each depending on ten extinctions, is 3.29. Spectrum III a !!! (Vogel.)

No. 63. Σ 3127, Dist. 18″. Observed as one mass. Birm., No. 403. The colour is not salient. The determination 1882.387 (see Memoirs R.A.S., vol. xlvii, p. 436) is rejected as erroneous.

No. 64. Birm., No. 404. The star is reddish.

No. 65. OΣ 328, Dist. 4″.4. Observed as one mass. Birm., No. 405. The red colour is not salient. Variability discovered by Schmidt, from 4.6 to 5.4 mag. Period 37 to 40 days.

No. 66. Suspected to be variable. A repetition of the determination of this star made on 1885.333 gave 4.86 as the magnitude. H.P. = 4.94.

No. 70. Spectrum III a ! (Vogel.)

No. 75. 5.40 is the mean of two determinations (5.43, 5.37) made on 1884.543 with twenty extinctions, and on 1885.352 with six. H.P. = 5.77.

No. 77. Σ 2161, Dist. 3″.7 Observed as one mass.

Reference Number.	Star's Designation.	R.A. 1890.	N.P.D. 1890.	Adopted Zenithal Magnitude. Polaris 2.05.	Average Deviation in Magnitude.	Date, 1880 +	Mag. Argel. Uran.
		h. m.	° ′				
81	77	17 23.8	41 39	5.89	0.05	4.543	6
82	λ	17 26.3	63 48	4.70	.09	...	5
83	P. xvii. 143	17 26.8	58 46	6.04	.02	4.582	6
84	78	17 27.5	61 31	5.71	.04	4.566	6
85	B.A.C. 5944	17 29.6	48 41	5.87	.03	4.582	6
86	P. xvii. 163	17 31.2	68 55	6.00	0.07	4.582	6
87	P. xvii. 176	17 32.4	59 9	5.73	.10	4.590	6
88	79	17 33.0	65 38	6.04	.08	4.566	6
89	82	17 33.7	41 21	5.45	.11	4.549	6
90	P. xvii. 196	17 35.8	58 44	6.45	.05	4.582	6
91	ι	17 36.4	43 56	4.11	0.12	2.376	3...4
92	84	17 38.9	65 37	6.19	.02	4.566	6
93	W.B. xvii. 1304...	17 41.4	58 27	6.49	.05	4.582	6
94	W.B. xvii. 1324...	17 42.1	72 13	5.62	.07	4.590	6
95	μ	17 42.2	62 12	3.50	.05	2.387	3...4
96	W.B. xvii. 1334...	17 42.2	51 4	6.17	0.03	4.582	6
97	D.M.+20°, No. 3570	17 43.6	69 24	5.96	.09	4.543	6
98	P. xvii. 255	17 44.0	70 42	6.04	.06	4.590	6
99	B.A.C. 6036	17 44.2	42 21	6.10	.06	4.546	6
100	87	17 44.3	64 20	5.73	.05	...	6
101	W.B. xvii. 1438...	17 46.0	60 38	6.17	0.04	...	6
102	W.B. xvii. 1433...	17 46.1	67 40	6.06	.02	4.566	6
103	88	17 47.2	41 35	6.41	.11	4.582	6
104	O.A. 17589	17 49.1	43 18	6.36	.07	4.582	6
105	90	17 49.7	49 58	5.20	.10	4.566	5

No. 82. 4.70 is the mean of two determinations (4.72, 4.69) made on 1882.387 with twenty extinctions, and on 1885.333 with six. **H.P.=4.29.** Birm., No. 411. The red colour is not salient.

No. 86. Σ 2190, Dist. 10″. Observed as one mass.

No. 90. Birm., No. 416. The star is slightly orange.

No. 94. Σ 2215, Dist. <1″. Observed as one mass. Spectrum I a! (Vogel.)

No. 95. Σ 2220, Dist. 30″. The faint companion not observed.

No. 98. Spectrum I a! (Vogel.)

No. 100. 5.73 is the mean of two determinations (5.71, 5.75) made on 1884.543 with twenty extinctions, and on 1885.333 with six. **H.P.=5.39.**

No. 101. 6.17 is the mean of two determinations (6.22, 6.12) made on 1884.549 with twenty extinctions, and on 1885.352 with six. **H.P.=5.66.**

No. 105. Suspected to be variable by Peirce. (See Photometric Researches, p. 149.)

Refer-ence Number.	Star's Designation.	R.A. 1890.	N.P.D. 1890.	Adopted Zenith'd Magnitude. Polaris 2.05.	Average Deviation in Magnitude.	Date, 1880 +	Mag. Argel. Uran.
		h. m.	o '				
106	89	17 51.1	63 56	5.87	0.07	4.590	6
107	W.B. xvii. 1591...	17 51.1	67 32	5.80	.06	4.582	6
108	θ	17 52.5	52 44	3.68	.07	...	4
109	ξ	17 53.5	60 44	3.99	.11	2.387	4...3
110	ν.................	17 54.3	59 48	4.58	.07	2.398	4...5
111	93	17 55.2	73 15	4.47	0.08	4.526	5
112	B.+45°, No. 2635	17 55.8	44 31	6.24	.06	4.582	6
113	B.+33°, No. 3006	17 56.5	56 47	5.00	.04	4.590	6
114	95	17 56.8	68 24	4.36	.09	...	4...5
115	B.A.C. 6109	17 56.9	44 29	5.72	.04	4.582	6
116	B.+33°, No. 3009	17 57.5	56 40	6.32	0.06	4.590	...
117	96	17 57.7	69 10	5.19	.10	4.546	5
118	B.A.C. 6129	18 0.3	41 32	6.16	.07	4.582	6
119	98	18 1.4	67 47	5.68	.09	...	6
120	W.B. xvii. 1941...	18 1.7	57 46	6.14	.02	5.352	6
121	99	18 2.9	59 27	5.21	0.07	4.582	5
122	ο.................	18 3.3	61 15	3.67	.11	2.398	4...3
123	100 1st star	18 3.4	63 55	6.02	.03	4.590	} 5...6
124	100 2nd star	18 3.6	63 55	6.16	.09	4.590	
125	102.................	18 4.0	69 12	4.26	·09	...	4...5
126	101.............:....	18 4.1	69 58	5.28	0.04	4.592	5
127	W.B. xviii. 76 ...	18 4.2	53 37	5.91	.05	...	6
128	B.A.C. 6162	18 4.2	46 33	4.06	.06	4.582	5
129	O.A. 17898	18 4.4	40 19	6.11	.09	4.592	6
130	W.B. xviii. 137...	18 6.1	53 34	6.09	.04	4.592	6

No. 108. 3.68 is the mean of two determinations (3.66, 3.69) made on 1882.387 with twenty extinctions, and on 1885.352 with six.

No. 111. Spectrum II a! (Vogel.)

No. 114. Σ 2264, Dist. 6″. Observed as one mass. 4.36 is the mean of two determinations (4.25, 4.47) made on 1882.398 with twenty extinctions, and on 1885 333 with six.

No. 118. Σ 2277, Dist. 28″. The bright star observed.

No. 119. 5.68 is the mean of two determinations (5.65, 5.71) made with twenty extinctions on 1884.593, and on 1885.333 with six. H.P. = 5.38.

No. 122. A strong suspicion of the variability of this star has been entertained by Schwab, Oudemans, and others. A second determination on 1885.352 with six extinctions gave 4.01 as the magnitude. H.P. = 3.99.

Nos. 123 and 124. These stars form Σ 2280, Dist. 14″.

No. 125. 4.26 is the mean of two determinations (4.16, 4.36) made on 1882.398 with twenty extinctions, and on 1885.333 with six.

No. 127. 5.91 is the mean of two determinations (5.89, 5.92) made on 1884.592 with twenty extinctions, and on 1885 352 with six. H.P. = 5.04.

Reference Number.	Star's Designation.	R.A. 1890.	N.P.D. 1890.	Adopted Zenithal Magnitude. Polaris 2.05.	Average Deviation in Magnitude.	Date, 1880 +	Mag. Argel. Uran.
		h. m.	° ′				
131	W.B. xviii. 183 ...	18 7.7	56 34	6.04	0.02	5.352	6
132	104..................	18 7.8	58 37	5.07	.05	4.582	5
133	Lal. 33610.........	18 8.8	41 43	6.49	.04	4.595	6
134	W.B. xviii. 319 ...	18 13.1	71 55	6.25	.11	4.595	6
135	105..................	18 14.7	65 36	5.53	.10	4.592	6
136	106..................	18 15.7	68 5	5.00	0.03	4.546	6
137	107..................	18 16.7	61 11	5.25	.16	4.582	5
138	108..................	18 16.7	60 12	5.72	.06	4.592	5...6
139	B.A.C. 6241	18 17.6	66 46	5.65	.06	4.595	6
140	B.A.C. 6245	18 18.1	72 14	5.41	.07	4.590	6
141	109..................	18 19.0	68 17	4.16	0.10	5.352	4
142	P. xviii. 83.........	18 22.0	63 37	6.83	.02	4.592	} 6
143	P. xviii. 84.........	18 22.2	63 37	6.01	.05	4.592	
144	P. xviii. 100	18 25.0	66 12	5.73	.05	4.595	6
145	W.B. xviii. 703 ...	18 26.0	73 9	5.80	.07	4.595	6
146	P. xviii. 116	18 28.2	66 28	5.86	0.06	4.592	6
147	W.B. xviii. 845 ...	18 30.2	71 54	5.69	.07	4.595	6
148	P. xviii. 132	18 30.9	66 29	5.71	.12	4.592	6
149	W.B. xviii. 906 ...	18 32.1	73 54	6.31	.08	4.595	6
150	B.A.C. 6341	18 39.9	66 33	6.40	.05	4.595	6
151	110..................	18 40.9	69 33	4.04	0.07	2.398	4
152	111..................	18 42.2	71 57	4.25	.04	2.398	4...5
153	B.+23°, No. 3461	18 43.5	66 36	6.22	.15	4.595	6
154	Lal. 35032.........	18 43.9	70 47	6.23	.06	4.592	6
155	112..................	18 47.6	68 42	5.46	.12	4.592	5
156	113..................	18 50.1	67 30	4.77	0.06	2.398	4...5
157	W.B. xviii. 1528	18 51.1	72 1	5.97	.10	4.595	6

No. 132.   Birm., No. 430.   The star is orange.
No. 134.   Spectrum I a!   (Vogel.)
No. 141.   The determination of 1882.398 (see Memoirs R.A.S., vol. xlvii, p. 437) is rejected. The observer's note is 'doubtful night, clouds and mist.'   Birm., No. 440.   The red colour is not salient.
Nos. 145 and 147.   Spectrum I a!   (Vogel.)
No. 148.   Birm., No. 452.   The red colour is not salient.
No. 152.   See note to No. 141.   A second determination with six extinctions made on 1885.352 gave 4.39.   H.P. – 4.52.   Spectrum I a!!   (Vogel.)
No. 154.   The determination was exactly repeated on 1885.352 with six extinctions. H.P. – 5.92.   Spectrum I a!   (Vogel.)
No. 157.   Birm., No. 472.   The red colour is not salient.

HYDRA.

Reference Number.	Star's Designation.	R.A. 1890.	N.P.D. 1890.	Adopted Zenithal Magnitude. Polaris 2.05.	Average Deviation in Magnitude.	Date. 1880 +	Mag. Argel. Uran.
		h. m.	° ′				
1	δ....................	8 31.8	83 55	4.23	0.15	...	4...5
2	σ	8 33.0	86 16	4.42	.06	4.209	5
3	η....................	8 37.5	86 12	4.42	.08	...	5...4
4	ε....................	8 41.0	83 11	3.55	.10	...	3...4
5	P. viii. 167.........	8 41.7	91 30	5.15	.07	4.220	5
6	ρ....................	8 42.6	83 45	4.66	0.11	...	5
7	14	8 43.8	93 2	5.07	.06	4.220	6
8	15	8 46.2	96 46	5.52	.12	4.215	6
9	ζ....................	8 49.6	83 38	3.42	.10	...	3...4
10	L. 17835	8 56.4	90 3	6.00	.08	...	6
11	ω	9 0.2	84 28	5.57	0.05	4.220	6
12	20	9 4.2	98 20	5.03	.05	4.229	6
13	21	9 7.0	96 40	6.20	.09	4.229	6
14	θ	9 8.6	87 13	4.03	.07	2.300	4
15	23	9 11.3	95 54	5.64	.12	4.215	6
16	24	9 11.3	98 17	5.40	0.03	4.229	6
17	27	9 15.1	99 5	5.01	.06	4.229	6
18	28	9 19.9	94 39	5.72	.12	4.215	6
19	α	9 22.2	98 11	2.22	.04	...	2
20	B.A.C. 3226	9 22.3	95 35	5.39	.04	4.231	6

No. 1. **4.23** is the mean of two determinations (4.09, 4.37) made on 1882.300 with twenty extinctions, and on 1883.207 with ten. Spectrum I a ! (Vogel.)

No. 2. Spectrum II a ! (Vogel.)

No. 3. **4.42** is the mean of two determinations (4.47, 4.37) made on 1883.300 with twenty extinctions, and on 1885.206 with six. H.P. = 4.17.

No. 4. Σ 1273, Dist. 3″.4. Binary. Observed as one mass. **3.55** is the mean of two determinations (3.58, 3.51) made on 1882.300 with twenty extinctions, and on 1883.207 with ten.

No. 6. **4.66** is the mean of two determinations (4.63, 4.69) made on 1884.209 with twenty extinctions, and on 1885.266 with six. H.P. = 4.31. Spectrum I a ! (Vogel.)

No. 9. **3.42** is the mean of two determinations (3.44, 3.40) made on 1882.300 with twenty extinctions, and on 1883.207 with ten.

No. 10. **6.00** is the mean of two determinations (5.93, 6.07) made on 1884.229 with twenty extinctions, and on 1885.266 with six. H.P. = 5.01.

No. 11. Called red in Uranometria Argentina.

No. 14. The faint distant companion not observed.

No. 19. Birm., No. 223. The colour is reddish. **2.22** is the mean of six accordant determinations, made on as many nights; two at Oxford and four at Cairo, involving sixty extinctions between 1883.094 and 1883.207.

Refer-ence Number.	Star's Designation.	R.A. 1890.	N.P.D. 1890.	Adopted Zenithal Magnitude. Polaris 2.05.	Average Deviation in Magnitude.	Date, 1880 +	Mag. Argel. Uran.
		h. m.	o '				
21	τ^1	9 23.6	92 17	4.99	0.02	4.209	5
22	τ^2	9 26.4	90 42	4.81	.04	4.231	5
23	33	9 29.1	95 25	5.76	.09	4.215	6
24	ι.....................	9 34.2	90 39	4.36	.03	4.231	4...5
25	L. 20850	10 44.2	99 16	6.45	.05	4.220	6

No. 21. The distant companion not observed.
No. 24. Birm., No. 226. The red colour is not salient. Spectrum II a !! (Vogel.)

LACERTA.

1	B.A.C. 7681	21 58.5	45 53	5.75	0.09	5.015	6
2	P. xxi. 405	22 1.5	45 31	5.31	.03	5.015	6
3	B.A.C. 7746	22 6.9	39 43	5.65	.02	5.025	5...6
4	P. xxii. 36	22 9.2	50 50	4.86	.01	5.031	5
5	R. 5614	22 9.2	45 6	5.96	.04	5.025	6
6	1....................	22 11.2	52 48	4.64	0.06	...	5...4
7	2....................	22 16.5	44 1	4.65	.06	2.675	5...4
8	3....................	22 19.2	38 19	4.50	.05	2.675	4.·5
9	4....................	22 20.0	41 5	4.86	.12	5.031	5
10	W.B. xxii. 467 ...	22 22.6	50 45	6.02	.05	5.025	6
11	5....................	22 24.9	42 51	4.48	0.09	5.031	5
12	6....................	22 25.7	47 26	4.72	.02	5.017	5
13	7....................	22 26.8	40 17	4.15	.04	2.675	4
14	B.A.C. 7858	22 27.6	50 47	5.80	.01	5.017	6
15	P. xxii. 163	22 31.0	50 57	5.67	.01	5.031	} 6
16	8....................	22 31.0	50 56	5.31	0.09	5.031	}
17	9....................	22 32.9	39 1	4.90	.02	5.017	5
18	10	22 34.3	51 31	5.18	.05	5.025	5
19	L. 44342	22 34.6	52 59	5.90	.06	5.031	6
20	11	22 35.7	46 18	4.80	.07	5.017	5

No. 1. B.A.C. assigns this star to Cygnus.
No. 4. Birm., No. 607. The colour is orange. Secchi suggests variability. (See Prodromo.)
No. 6. Birm., No. 608. The star is slightly red. 4.64 is the mean of two determinations (4.67, 4.60) made on 1882.675 with twenty extinctions, and on 1885.485 with ten. H.P. = 4.14.
No. 8. Birm., No. 611. The red colour is not salient.
No. 11. Birm., No. 612. The star is of a deep orange colour.
Nos. 15 and 16. These stars form Σ 2922, Dist. 22".5.
No. 20. Birm., No. 616. The red colour is not salient.

Reference Number.	Star's Designation.	R.A. 1890.	N.P.D. 1890.	Adopted Zenithal Magnitude. Polaris 2.05.	Average Deviation in Magnitude.	Date, 1880 +	Mag. Argel. Uran.
		h. m.	° ′				
21	12	22 36.5	50 21	5.28	0.01	5.025	6
22	13	22 39.2	48 45	5.14	.01	5.031	6
23	14	22 45.4	48 38	5.80	.10	5.017	6
24	15	22 47.1	47 16	5.28	.07	5.025	6
25	B.A.C. 7983	22 48.8	45 50	5.75	.10	5.031	6
26	B.A.C. 7984	22 49.0	50 13	5.96	0.01	5.031	6
27	W.B. xxii. 1121...	22 49.9	53 30	5.86	.04	5.025	6
28	W.B. xxii. 1133...	22 50.6	54 13	6.06	.02	...	6
29	16	22 51.4	48 59	5.84	.01	5.031	6
30	B.A.C. 7995	22 51.6	40 51	5.22	.02	5.025	6
31	B.A.C. 7999	22 52.2	41 54	5.30	0.07	5.031	6
32	R. 5895	22 52.5	51 15	6.11	.01	5.025	6

No. 24. Birm., No. 622. The red colour is not salient.
No. 28. 6.06 is the mean of two determinations (6.01, 6.12) made on 1885.025 and 1885.485, each with ten extinctions. H.P. = 5.60.
No. 29. Σ 2960, Dist. 64″. The larger star observed. Birm., No. 624. The red colour is not salient. Variable ?

LEO.

1	κ...............	9 18.2	63 21	4.69	0.05	5.322	5
2	ω	9 22.6	80 28	5.53	.05	4.327	6
3	3	9 22.6	81 20	6.05	.04	4.330	6
4	λ	9 25.5	66 33	4.49	.07	2.357	5...4
5	ξ	9 26.0	78 13	5.22	.03	4.330	6
6	6	9 26.1	79 48	5.55	0.04	4.327	6
7	P. ix. 124	9 30.2	58 21	5.66	.03	4.330	6
8	8...............	9 31.0	73 4	6.18	.05	4.330	6
9	P. ix. 145	9 35.1	58 13	6.30	.05	4.338	6
10	o	9 35.3	79 36	3.75	.08	2.311	4...3

No. 1. The determination of 1882.357 (see Memoirs R.A.S., vol. xlvii, p. 438) is rejected as apparently erroneous.
No. 2. Σ 1356, Dist. 1″. Binary. Period 111 years. Observed as one mass.
No. 4. Birm., No. 224. The star is slightly red.
No. 6. Σ Unnum., Dist. 38″. The faint companion not observed. Birm., Add. I, No. 41. The colour is yellow.

Reference Number.	Star's Designation.	R.A. 1890.	N.P.D. 1890.	Adopted Zenithal Magnitude. Polaris 2.05.	Average Deviation in Magnitude.	Date, 1880+	Mag. Argel. Uran.
		h. m.	o '				
11	15	9 37.1	59 31	5.69	0.04	4.330	5
12	ψ	9 37.7	75 29	5.77	.03	4.338	6
13	ε	9 39.6	65 43	3.39	.04	2.344	3
14	18	9 40.5	77 41	6.06	.04	4.327	6
15	22	9 45.6	65 5	5.26	.04	4.330	5
16	μ	9 46.5	63 28	3.92	0.05	2.344	4
17	ν	9 52.3	77 2	5.38	.04	4.330	5
18	P. ix. 221	9 53.3	59 50	6.08	.06	4.330	6
19	π	9 54.4	81 26	5.17	.08	4.327	5
20	P. ix. 230	9 56.7	67 31	5.69	.06	4.330	6
21	η	10 1.3	72 42	3.48	0.16		3...4
22	31	10 2.1	79 28	4.87	.08		5
23	α	10 2.5	77 30	1.17	.07		1...2
24	ζ	10 10.6	66 2	3.39	.11		3
25	γ	10 13.9	69 36	2.12	.07		2
26	44	10 19.5	80 39	6.23	0.08	4 327	6
27	45	10 21.8	79 41	5.79	.07	4.338	6
28	46	10 26.3	75 18	5.76	.04	5.322	6
29	ρ................	10 27.0	80 8	3.96	.10	...	4
30	48	10 29.1	82 29	5.02	.06	4.327	6
31	50	10 33.0	73 18	6.38	0.07	4 327	6
32	51	10 40.5	70 32	5.06	.05	4.344	6
33	52	10 40.7	75 13	5.75	.04	5.322	6
34	53	10 43.5	78 52	5.40	.05	4 346	5
35	μ¹ (1st star)	10 48.1	91 33	6.16	.09	4.346	6

No. 19. Birm., No. 231. The red colour is not salient. Spectrum III a!! (Vogel.)

No. 21. **3.48** is the mean of two determinations (3.48, 3.47) made on 1882.344 with twenty extinctions, and on 1883.097 with ten.

No. 22. Birm., No. 233. The colour is slightly orange. Spectrum II a!! (Vogel.) **4.87** is the mean of two determinations (4.84, 4.90) made on 1882.311 with twenty extinctions, and on 1883.097 with ten.

No. 23. Σ 6ᴵᴵ. Spectrum I a!!! (Vogel.) **1.17** is the mean of seven determinations, made on as many nights, four at Oxford and three at Cairo, involving eighty extinctions, between 1882.311 and 1883.127. H.P.=1.42.

No. 24. Σ 18¹. The distant companion 314″ not observed. **3.39** is the mean of two determinations (3.37, 3.42) made on 1882.344 with twenty extinctions, and on 1883.097 with ten.

No. 25. Σ 1424, Dist. 3″. Binary. Period 407 years. Birm., No. 237. The red colour is not salient. **2.12** is the mean of two determinations (2.08, 2.16) made on 1882.357 with twenty extinctions, and on 1883.097 with ten.

No. 28. Spectrum III a! (Vogel.)

No. 29. **3.96** is the mean of two determinations (3.94, 3.98) made on 1882.344 with twenty extinctions, and on 1883.097 with ten.

No. 34. Spectrum I a! (Vogel.)

Reference Number.	Star's Designation.	R.A. 1890.	N.P.D. 1890.	Adopted Zenithal Magnitude. Polaris 2.05.	Average Deviation in Magnitude.	Date, 1880 +	Mag. Argel. Uran.
		h. m.	° ′				
36	p^1 (2nd star)	10 48.1	91 33	6.00	0.11	4.346	6
37	54	10 49.7	64 40	4.46	.09	2.344	4...5
38	55	10 50.1	88 41	6.20	.04	4.327	6
39	P. x. 194	10 50.4	67 4	6.32	.06	4.338	6
40	58	10 54.9	85 47	5.01	.06	5.322	5
41	59	10 55.1	83 18	5.13	0.03	4.330	5
42	61	10 56.2	91 53	5.06	.04	4.344	5
43	60	10 56.4	69 13	4.56	.07	2.357	4...5
44	62	10 58.0	89 24	6.26	.02	4.338	6
45	χ	10 59.4	82 4	4.98	.08	...	5
46	65	11 1.3	87 27	5.81	0.03	4.330	6
47	67	11 2.9	64 45	5.74	.08	4.344	6
48	69	11 8.1	89 28	5.52	.04	4.344	5
49	δ	11 8.3	68 52	2.55	.03	2.357	2...3
50	P. xi. 12	11 8.3	81 20	6.09	.01	4.346	6
51	θ...	11 8.5	73 58	3.44	0.07	2.357	3.. 4
52	72	11 9.4	66 18	5.93	.02	4.376	5
53	73	11 10.1	76 6	5.74	.05	4.344	6
54	φ	11 11.1	93 3	4.31	.04	2.360	5...4
55	75	11 11.6	87 23	5.63	.05	4.327	6
56	σ	11 15.5	83 22	4.15	0.08	...	4
57	ι	11 18.2	78 52	4.18	.05	2.360	4
58	79	11 18.4	87 59	5.45	.04	4.327	6
59	P. xi. 60	11 19.3	77 58	6.03	.06	4.344	6
60	81	11 19.9	72 56	5.77	.03	...	6

No. 37. Σ 1487, Dist. 6″. Observed as one mass.

No. 40. The observation of 1882.357 (see Memoirs R.A.S., vol. xlvii, p. 439) is rejected. See note to No. 1, Leo. Spectrum II a! (Vogel.)

No. 41. Spectrum I a! (Vogel.)

No. 42. Spectrum III a! (Vogel.) Called red in the Uranometria Argentina.

No. 43. Birm., No. 251. The red colour is not salient.

No. 45. 4.98 is the mean of two determinations (5.00, 4.96) made on 1882.357 with twenty extinctions, and on 1883.097 with ten.

No. 47. = 53 Leonis Minoris. (See Introduction to B.A.C., p. 75.)

No. 48. Spectrum I a!! (Vogel.) No. 51. Spectrum I a!!! (Vogel.)

No. 52. Birm., No. 256. The red colour is not salient.

No. 54. Gould suspects variability. See Uranometria Argentina, p. 337.

No. 55. Spectrum III a! (Vogel.)

No. 56. 4.15 is the result of two identical determinations made on 1882.360 with twenty extinctions, and on 1883.097 with ten.

No. 57. Σ 1536, Dist. 3″. Observed as one mass.

No. 60. 5.77 is the mean of two determinations (5.73, 5.80) made on 1884.338 with twenty extinctions, and on 1885.322 with six. H.P. = 5.47.

Reference Number.	Star's Designation.	R.A. 1890.	N.P.D. 1890.	Adopted Zenithal Magnitude. Polaris 2.05.	Average Deviation in Magnitude.	Date, 1880 +	Mag. Argel. Uran.
		h. m.	o ′				
61	τ....................	11 22.3	86 32	5.18	0.06	4.327	5
62	85	11 24.0	73 59	6.00	.08	4.346	6
63	86	11 24.8	70 59	6.03	.09	...	6
64	87	11 24.7	92 24	5.00	.04	4.344	5
65	88	11 26.1	75 1	6.25	.03	4.344	6
66	89	11 28.8	86 20	6.09	0.04	...	6
67	90	11 29.0	72 36	5.90	.05	4.376	6
68	L. 21984	11 30.0	78 29	6.48	.03	5.322	6
69	P. xi. 111	11 30.5	61 36	5.82	.05	4.327	6
70	υ....................	11 31.3	90 13	4.32	.06	4.346	5...4
71	92	11 35.1	68 2	5.25	0.06	4.344	5
72	93	11 42.3	69 10	4.39	.05	2.360	4...5
73	β	11 43.5	74 49	2.07	.02	...	2
74	B.A.C. 3997	11 43.6	73 9	5.89	.05	4.346	6
75	B.A.C. 4005	11 45.3	77 6	6.00	.02	4.344	6
76	95	11 50.0	73 44	5.51	0.03	4.346	6

No. 61. Σ 19¹, Dist. 95″. The bright star observed.
No. 62. Spectrum II a ! (Vogel.)
No. 63. 6.03 is the mean of two determinations (6.01, 6.04) made on 1884.346 with twenty extinctions, and on 1885.322 with six. H.P. = 5.72.
No. 64. Birm., No. 261. The red colour is not salient.
No. 65. Σ 1547, Dist. 15″. Observed as one mass.
No. 66. 6.09 is the mean of two determinations (6.07, 6.11) made on 1884.327 with twenty extinctions, and on 1885.322 with six. H.P. = 5.73.
No. 67. Σ 1552. Triple. The close pair 3″.5 distance, observed as one mass.
No. 68. Spectrum I a ! (Vogel.)
No. 69. Σ 1555, Dist. 1″. Observed as one mass. B.A.C. assigns this star to Ursa Major.
No. 70. Called red in Uranometria Argentina.
No. 73. Birm., No. 267. The red colour is not salient. Suspected to be variable by Sir W. Herschel and Schmidt. (See Ast. Nach., xxi, p. 110.) Spectrum I a !!! (Vogel.) 2.07 is the mean of five accordant determinations, made on as many nights, three at Oxford and two at Cairo, involving sixty extinctions, between 1882.360 and 1883.127. H.P. = 2.23.
No. 76. Spectrum I a ! (Vogel.)

LEO MINOR.

Reference Number.	Star's Designation.	R.A. 1890.	N.P.D. 1890.	Adopted Zenithal Magnitude. Polaris 2.05.	Average Deviation in Magnitude.	Date, 1880 +	Mag. Argel. Uran.
		h. m.	o ,				
1	8	9 24.9	54 25	5.46	0.05	4.314	6
2	10	9 27.5	53 7	4.65	.05	5.327	5
3	11	9 29.1	53 41	5.43	.06	4.311	6
4	19	9 51.0	48 25	5.18	.07	4.311	5
5	20	9 54.7	57 32	6.01	.08	4.314	5
6	21	10 1.0	54 13	4.42	0.08	2.368	4...5
7	22	10 8.8	57 59	6.47	.05	4.314	6
8	23	10 10.0	60 9	5.29	.07	4.311	5
9	27	10 16.8	55 32	6.08	.08	4.314	6
10	30	10 19.6	55 39	4.83	.04	5.327	5...4
11	31 (β)	10 21.6	52 44	4.18	0.03	2.368	4...5
12	32	10 23.7	50 31	5.67	.05	4.314	6
13	34	10 27.2	54 27	5.55	.08	4.314	6
14	35	10 30.1	53 6	6.18	.10	5.327	6
15	37	10 32.5	57 27	4.87	.08	2.368	5...4
16	38	10 32.9	51 31	5.86	0.04	4.311	6
17	40	10 37.0	63 6	5.39	.06	4.293	6
18	41	10 37.4	66 14	5.12	.05	4.311	5
19	42	10 39.8	58 44	5.23	.08	4.308	5
20	46	10 47.2	55 11	4.01	.08	2.368	4
21	47	10 48.9	55 23	6.01	0.03	4.314	6

Nos. 2 and 10. The determinations of 1882.368 (see Memoirs of R.A.S., vol. xlvii, p. 440) are rejected. The meteorological conditions were variable.
No. 18. The B.A.C. assigns this star to Leo.
No. 21. Identification of Argelander's star doubtful.

LYNX.

Reference Number.	Star's Designation.	R.A. 1890.	N.P.D. 1890.	Adopted Zenithal Magnitude. Polaris 2.05.	Average Deviation in Magnitude.	Date, 1880+	Mag. Argel. Uran.
		h. m.	° ′				
1	1	6 7.8	28 27	5.42	0.08	5.247	6
2	2	6 9.9	30 57	4.73	.07	...	5...4
3	B.A.C. 2046	6 17.2	33 39	5.52	.01	5.247	6
4	5	6 17.2	31 31	5.61	.09	5.301	6
5	8	6 27.6	28 25	6.08	.01	5.280	6
6	11	6 28.3	33 3	5.86	0.12	5.247	6
7	12	6 36.5	30 26	4.85	.05	5.250	5
8	13	6 37.5	32 43	5.63	.07	5.193	6
9	R. 1806	6 39.0	34 10	5.66	.04	5.280	6
10	14	6 43.4	30 26	5.57	.02	5.247	6
11	15	6 47.8	31 26	4.70	0.01	5.247	5
12	18	7 6.3	30 10	5.25	.04	5.280	6
13	B.A.C. 2361	7 7.7	42 33	5.76	.05	5.247	6
14	B.A.C. 2367	7 8.9	37 40	6.02	.09	5.280	6
15	B.A.C. 2379	7 10.2	40 20	4.91	.01	5.223	6
16	19	7 13.9	34 31	5.18	0.03	5.247	5
17	21	7 18.4	40 34	4.88	.04	...	5
18	R. 1960	7 20.7	41 35	5.68	.10	5.280	6
19	22	7 21.6	40 6	5.57	.01	5.247	6
20	B.A.C. 2488	7 28.5	43 35	5.61	.10	5.223	6
21	R. 2005	7 33.1	41 36	5.76	0.02	5.280	6
22	24	7 33.7	31 2	5.10	.06	5.247	5
23	P. vii. 169	7 35.7	39 18	5.50	.05	...	6
24	W.B. vii. 1083 ...	7 39.3	52 13	5.59	.06	5.193	6
25	25	7 46.5	42 20	6.14	.06	5.250	6

No. 2. 4.73 is the mean of two determinations (4.83, 4.63) made on 1882.371 with twenty extinctions, and on 1885.301 with six. H.P.=4.33.

No. 7. Σ 948, Triple. Distances 1″.4 and 8″.5 respectively. Observed as one mass.

No. 9. Σ 958, Dist. 5″. Observed as one mass.

No. 10. Σ 963, Dist. <1″. Observed as one mass.

No. 16. Σ 1062. Triple. The closer components, Dist. 14″, observed as one mass.

No. 17. 4.88 is the mean of two determinations (4.83, 4.94) made on 1885.250 and on 1885.301, each with six extinctions. H.P.=4.57.

No. 23. 5.50 is the mean of two determinations (5.57, 5.44) made on 1885.247 and on 1885.301, each with six extinctions. H.P.=5.20.

Reference Number.	Star's Designation.	R.A. 1890.	N.P.D. 1890.	Adopted Zenithal Magnitude. Polaris 2.05.	Average Deviation in Magnitude.	Date, 1880+	Mag. Argel. Uran.
		h. m.	o ,				
26	26	7 46.7	42 9	5.66	0.07	5.250	6
27	27	8 0.2	38 10	4.99	.04	2.371	5...4
28	B.A.C. 2704	8 1.1	31 26	5.04	.03	5.283	6
29	R. 2101	8 4.1	33 13	5.61	.11	5.280	6...5
30	29	8 8.7	30 5	5.52	.05	5.247	6
31	30	8 11.6	31 55	5.80	0.02	5.283	6
32	31	8 15.2	46 27	4.58	.06	5.280	5
33	P. viii. 40	8 15.5	36 25	5.07	.12	5.283	6
34	33	8 27.7	53 12	5.72	.01	5.193	6
35	34	8 33.4	43 47	5.60	.10	5.280	6
36	35	8 44.5	45 52	5.41	0.03	5.283	6
37	P. viii. 202.........	8 49.4	43 57	6.10	.05	5.280	6
38	36	9 6.6	46 20	5.10	.06	5.193	5
39	38	9 12.0	52 43	3.89	.10	2.371	4
40	40	9 14.4	55 8	3.16	.07	...	3...4
41	P. ix. 115	9 28.2	49 53	4.91	0.01	5.193	5
42	42	9 31.5	49 16	5.33	.13	5.193	6
43	43	9 35.2	49 44	5.50	.02	5.193	6

No. 28. B.A.C. assigns this star to Camelopardalus. Birm., No. 196. The red colour is not salient.

No. 31. B.A.C. assigns this star to Camelopardalus.

No. 39. Σ 1334, Dist. 3″. Observed as one mass.

No. 40. Σ 1342. The faint companion, dist. 18″, not observed. Birm., No. 220. The colour is orange. 3.16 is the mean of two determinations (3.13, 3.19) made on 1882.371 with twenty extinctions, and on 1885.301 with six.

LYRA.

Reference Number.	Star's Designation.	R.A. 1890.	N.P.D. 1890.	Adopted Zenithal Magnitude. Polaris 2.05.	Average Deviation in Magnitude.	Date, 1880+	Mag. Argel. Uran.
		h. m.	° ′				
1	B.A.C. 6193	18 9.4	51 16	6.01	0.07	4.756	6
2	B.A.C. 6203	18 12.2	47 53	5.66	.03	...	6...5
3	B.A.C. 6218	18 13.6	49 7	5.96	.06	4.756	6
4	κ	18 16.0	53 59	4.54	.04	2.656	5...4
5	μ	18 20.6	50 33	5.15	.05	4.746	5...6
6	L. 34132	18 21.8	60 14	5.78	0.08	4.768	6...5
7	W.B. xviii. 794 ...	18 28.7	59 31	5.62	.04	...	6...5
8	W.B. xviii. 894 ...	18 31.3	55 38	5.85	.06	4.780	6
9	W.B. xviii. 934 ...	18 32.6	56 37	5.52	.06	4.756	6...5
10	a	18 33.2	51 19	+ 0.86	.05	...	1
11	L. 34853	18 39.8	58 11	5.72	0.06	4.768	6
12	ε...................	18 40.7	50 27	4.39	.03	2.497	} 4
13	5	18 40.7	50 30	4.75	.12	2.497	
14	{ ζ (6)	18 41.0	52 29	4.54	.06	2.497	} 4 ..5
	7	18 41.1	52 30	6.02	.05	4.756	
15	W.B. xviii. 1218...	18 41.6	63 27	5.01	.08	4.770	5
16	B.A.C. 6404	18 42.7	48 41	5.80	0.07	4.746	6
17	L. 35045	18 43.8	58 22	5.86	.06	4.770	6
18	8...................	18 45.7	57 19	5.71	.05	4.768	...
19	ν...................	18 45.8	57 35	5.31	.05	4.768	5...6
20	W.B. xviii. 1460	18 48.6	48 45	6.02	.06	4.756	6
21	δ	18 50.7	53 14	4.68	0.06	2.497	4 .5
22	B.A.C. 6468	18 50.8	56 12	6.17	.05	4.770	6
23	B.A.C. 6473	18 51.3	48 33	5.62	.03	4.746	6
24	13	18 52.0	46 12	4.64	.03	4.768	5...4
25	D.M.+38°, No. 3373	18 54.3	51 53	5.80	.07	4.756	6

No. 2. **5.66** is the mean of two determinations (5.69, 5.64) made on 1884.768 with twenty extinctions, and on 1885.448 with six. H.P.= 5.34.

No. 7. **5.62** is the mean of two determinations (5.58, 5.67) made on 1884.770 with twenty extinctions, and on 1885.442 with six. H.P. = 5.39.

No. 9. Σ 2349, Dist. 7″. Observed as one mass.

No. 10. Σ 9ᴵᴵ. Observed as one mass. + 0.86 is the mean of six determinations made on as many nights, two at Oxford and four at Cairo, involving seventy extinctions between 1882.680 and 1883.185. H.P. = + 0.81. For explanation of notation see Preface.

No. 12. Σ 2382, Dist. 3″. Observed as one mass.

No. 13. Σ 2383, Dist. 2″.5. Observed as one mass. No. 14. Σ 38ᴵ, Dist. 44″.

No. 21. Birm., No. 470. The colour is slightly orange.

No. 24. Variability discovered by Baxendell in 1856. Max. 4.3, Min. 4.6 mag. Period 46 days. Birm., No. 474. The red colour is not salient.

Reference Number.	Star's Designation.	R.A. 1890.	N.P.D. 1890.	Adopted Zenithal Magnitude. Polaris 2.05.	Average Deviation in Magnitude.	Date, 1880 +	Mag. Argel. Uran.
		h. m.	° ′				
26	γ	18 54.8	57 28	3.16	0.03	2.656	3...4
27	B.A.C. 6493	18 55.2	49 29	5.99	.13	4.746	6
28	W.B. xviii. 1670	18 55.3	63 56	5.46	.05	4.785	6
29	B.A.C. 6495	18 55.5	50 56	6.30	.04	4.785	6
30	λ	18 55.9	58 1	5.22	.05	4.756	5...6
31	W.B. xviii. 1721	18 56.8	63 52	5.56	0.05	4.785	6
32	P. xviii. 290	18 56.4	56 25	6.14	.09	4.768	6
33	16	18 58.3	43 13	5.20	.02	4.746	5
34	P. xviii. 318	19 2.3	61 33	5.48	.07	4.785	6...5
35	W.B. xix. 20	19 2.7	48 45	6.09	.07	4.756	6
36	17	19 3.3	57 40	5.11	0.06	4.785	5...6
37	ι	19 3.4	54 4	5.31	.12	4.746	5
38	Σ 2470	19 4.8	55 25	6.47	.03	4.785	} 6
39	Σ 2474	19 5.1	55 35	6.48	.04	4.785	
40	D.M.+26°, No. 3476	19 7.1	63 54	6.82	.04	4.803	
41	D.M.+26°, No. 3477	19 7.2	63 56	7.14	0.05	4.802	} 6
42	19	19 7.5	58 54	5.80	.11	4.746	6
43	L. 36283	19 9.9	53 46	6.60	.05	...	6
44	η	19 10.0	51 3	4.68	.04	2.497	4...5
45	W.B. xix. 304 ...	19 11 5	62 45	6.07	.04	4.768	6
46	D.M.+27°, No. 3314	19 11.6	62 16	6.11	0.06	4.768	...
47	θ	19 12.6	52 4	4.64	.02	...	4...5
48	W.B. xix. 437 ...	19 15.1	52 45	6.10	.06	4.802	6
49	W.B. xix. 457 ...	19 16.6	55 1	5.98	.03	4.768	6
50	W.B. xix. 512 ...	19 18.5	56 42	6.13	.04	4.802	6
51	B.A.C. 6656	19 20.5	46 49	5.79	0.09	4.746	6

No. 30. Birm., No. 479. The colour is orange.
No. 36. Σ 2461, Dist. 4″. Observed as one mass.
No. 38. Dist. 13″. Observed as one mass.
No. 39. Dist. 17″. Observed as one mass.
No. 43. 6.60 is the mean of two determinations (6.61, 6.58) made on 1884.779 and on 1885.448, each with six extinctions.
No. 44. Σ 2487, Dist. 28″. The faint companion not observed.
No. 47. 4.64 is the mean of two determinations (4.67, 4.61) made on 1882.497 with twenty extinctions, and on 1885.448 with six. H.P. = 4.34.
No. 51. B.A.C. assigns this star to Cygnus.

MONOCEROS.

Reference Number.	Star's Designation.	R.A. 1890.	N.P.D. 1890.	Adopted Zenithal Magnitude. Polaris 2.05.	Average Deviation in Magnitude.	Date, 1880 +	Mag. Argel. Uran.
		h. m.	° ′				
1	2......................	5 53.9	99 34	5.06	0.06	4.201	6...5
2	B.A.C. 1994	6 6.5	96 31	4.90	.03	4.190	6...5
3	L. 11916	6 9.2	94 32	6.06	.04	4.190	6
4	5	6 9.5	96 14	4.38	.04	2.971	5...4
5	L. 11927	6 10.0	85 41	6.70	.14	4.097	6
6	L. 11949	6 10.1	94 53	6.06	0.06	4.190	6
7	L. 12074	6 13.6	99 21	5.61	.04	4.196	6
8	7......................	6 14.4	97 47	5.16	.03	4.201	6
9	L. 12104	6 14.5	92 54	5.00	.04	4.196	6
10	L. 12146	6 15.7	87 41	6.01	.06	4.196	6
11	8......................	6 17.9	85 21	4.89	0.06	...	5...4
12	L. 12216	6 18.0	81 3	5.83	.04	4.218	6
13	D.M.+2°, No. 1237	6 21.6	87 1	6.00	.02	4.218	6
14	10	6 22.5	94 42	4.82	.05	4.204	5
15	11	6 23.5	96 58	4.32	.06	2.971	4...5
16	L. 12494	6 25.7	78 23	4.89	0.03	4.239	5
17	12	6 26.5	85 4	6.03	.04	...	5
18	L. 12545	6 26.6	98 5	5.43	.04	4.196	6
19	13	6 27.0	82 35	4.78	.04	5.206	5...4
20	L. 12552	6 27.0	95 47	5.82	.03	4.190	6...5

No. 4. Birm., No. 140. The colour is orange.

No. 9. An orange star, suspected of variability by Gould and others. (See Uranometria Argentina, p. 331.)

No. 11. Σ 900, Dist. 13″. Observed as one mass. 4.89 is the mean of six determinations made on as many nights, one at Oxford and five at Cairo, involving seventy extinctions, between 1882.971 and 1883.138. H.P.=4.40. Spectrum I a! (Vogel.)

No. 15. Σ 919, Distances 7″ and 10″. Triple. Observed as one mass.

No. 16. Spectrum I a! (Vogel.)

No. 17. 6.03 is the mean of two determinations (6.00, 6.05) made on 1884.196 with twenty extinctions, and on 1885.174 with six. H.P. = 5.69.

No. 18. Called red in Uranometria Argentina.

No. 19. The determination of 1882.971 is rejected as erroneous. (See Memoirs R.A.S., vol. xlvii, p. 441.)

Reference Number.	Star's Designation.	R.A. 1890.	N.P.D. 1890.	Adopted Zenithal Magnitude. Polaris 2.05.	Average Deviation in Magnitude.	Date, 1880 +	Mag. Argel. Uran.
		h. m.	o ′				
21	L. 12587	6 28.0	91 8	4.73	0.05	4.190	5...6
22	P. vi. 171	6 31.2	95 7	5.61	.03	4.201	6
23	L. 12699	6 31.5	83 46	5.92	.09	4.220	6
24	15	6 34.9	80 0	4.58	.08	4.229	4
25	L. 12907	6 36.7	99 4	5.91	.10	4.196	6
26	16	6 40.6	81 18	5.45	0.05	4.204	6
27	17	6 41.4	81 51	5.03	.03	4.196	5
28	18	6 42.1	87 28	4.96	.06	...	5
29	L. 13100	6 42.4	98 53	5.55	.05	4.201	6
30	L. 13216	6 45.4	97 55	6.13	.04	4.201	6
31	P. vi. 257	6 46.8	81 29	5.98	0.07	...	6
32	D.M.+10°, No. 1335	6 50.4	79 54	5.77	.01	4.229	6
33	W.B. vi. 1580 ...	6 53.4	82 32	6.29	.09	4.204	6
34	L. 13627	6 56.6	95 34	5.26	.09	4.201	6
35	B.A.C. 2304	6 57.3	80 42	6.12	.08	4.190	6
36	19	6 57.5	94 5	4.85	0.04	4.196	6
37	L. 13781	7 1.3	84 55	5.80	.01	4.220	6
38	L. 13799	7 1.9	82 21	6.06	.03	4.204	6
39	20	7 4.8	94 4	4.94	.04	4.190	6
40	22	7 6.3	90 19	4.46	.04	2.971	4...5
41	B.A.C. 2373	7 8.6	86 42	5.58	0.06	4.220	6
42	P. vii. 86	7 17.0	95 46	6.16	.03	4.201	6
43	25	7 31.8	93 52	4.05	.04	4.196	5 ..6
44	26	7 36.0	99 18	4.39	.10	2.971	4...5
45	L. 15136	7 40.7	96 30	5.98	.17	4.201	6

No. 24. Σ 950, Dist. 2″.5. Observed as one mass. Variability discovered by Winnecke in 1867. Max. 4.9, Min. 5.4. Period $3^d.4$.

No. 25. Birm., No. 151. The red colour is not salient.

No. 27. Birm., Add. II, No. 4. The colour is slightly orange.

No. 28. 4.96 is the mean of two determinations (5.07, 4.84) made on 1884.204 with twenty extinctions, and on 1885.206 with six. H.P. = 4.75.

No. 31. 5.98 is the mean of two determinations (5.96, 5.99) made on 1884.196 with twenty extinctions, and on 1885.206 with six. H.P. = 5.72.

No. 34. Birm., Add. I, No. 31. The star is reddish.

No. 36. Suspected to be variable. (See Uranometria Argentina, p. 332.)

No. 40. Spectrum I a! (Vogel.)

No. 41. B.A.C. assigns this star to Canis Minor.

No. 44. Birm., Add. II, No. 8. The red colour is not salient.

Reference Number.	Star's Designation.	R.A. 1890.	N.P.D. 1890.	Adopted Zenithal Magnitude. Polaris 2.05.	Average Deviation in Magnitude.	Date, 1880 +	Mag. Arael. Uran.
		h. m.	o ,				
46	P. vii. 228........	7 44.9	98 54	5.92	0.06	4.201	6
47	L. 15374	7 47.3	95 8	5.56	.08	4.201	6
48	27	7 54.2	93 23	5.18	.09	4.201	6...5
49	28	7 55.6	91 5	4.82	.09	4.196	5...6
50	L. 15717	7 57.0	96 2	6.03	.05	4.204	6
51	Σ 1183	8 1.2	98 56	5.96	0.04	4.204	6
52	29	8 3.1	92 40	5.32	.06	2.971	5
53	L. 16049	8 6.2	97 27	5.56	.04	4.204	6
54	30	8 20.2	93 33	3.63	.03	2.971	4...3
55	L. 16837	8 28.5	91 47	5.41	.09	4.204	6
56	31	8 38.3	96 50	4.60	0.04	4.204	5

No. 47. Variable? (See Uranometria Argentina, p. 333.)
No. 49. Birm., Add. II, No. 10. The red colour is not salient.
No. 51. Dist. 31″. The bright star observed.
No. 52. Σ 1190. Triple. The bright star observed.
Nos. 54 and 56. B.A.C. assigns these stars to Hydra.
No. 55. Spectrum I a! (Vogel.)

OPHIUCHUS.

1	δ	16 8.6	93 25	2.65	0.04	2.475	3
2	ε	16 12.5	94 25	3.28	.10	2.475	3...4
3	υ	16 21.9	98 8	4.64	.07	4.505	5
4	W.B. xvi. 394 ...	16 23 0	89 5	5.59	.03	4.498	6
5	λ	16 25.4	87 46	3.81	.09	2.475	4...3
6	12	16 30.6	92 5	5.80	0.06	4.508	6
7	ζ	16 31.1	100 21	2.78	.06	...	3...2
8	L. 30232	16 32.1	96 19	6.20	.04	4.498	6
9	L. 30351	16 35.9	88 33	7.13	.07	4.505	} 6
10	14	16 36.1	88 37	6.03	.04	4.505	

No. 1. Birm., No. 373. The colour is reddish.
No. 2. Birm., No. 375. The red colour is not salient.
No. 5. Σ 2055, Dist. 1″.5. Binary. Period 234 years. Observed as one mass. Spectrum I a! (Vogel.)
No. 7. 2.78 is the mean of four determinations made on as many nights at Cairo, involving forty extinctions between 1883.161 and 1883.171.

Refer-ence Number.	Star's Designation.	R.A. 1890.	N.P.D. 1890.	Adopted Zenithal Magnitude. Polaris 2.05.	Average Deviation in Magnitude.	Date, 1880 +	Mag. Argel. Uran.
		h. m.	° ′				
11	19	16 41.6	87 46	6.21	0.02	4.498	6
12	21	16 45.8	88 36	5.56	.05	5.363	6
13	23	16 48.7	95 58	5.65	.04	4.508	6
14	ι	16 48.8	79 39	4.16	.05	...	4...5
15	κ	16 52.3	80 28	3.17	.03	2.475	3...4
16	30	16 55.3	94 3	5.00	0.10	4.505	5
17	P. xvi. 289	16 59.9	90 45	5.68	.04	4.498	6
18	η	17 4.1	105 35	...	.09	2.475	2...3
19	37	17 7.3	79 17	5.57	.03	5.360	6
20	W.B. xvii. 143 ...	17 10.9	88 40	5.68	.06	4.522	6
21	41	17 11.0	90 19	5.14	0.04	4.519	5
22	(e) B.A.C. 5841...	17 13.5	79 1	5.45	.06	4.508	5
23	θ	17 15.3	114 53	...	.03	2.475	3...4
24	P. xvii. 84	17 18.7	81 3	6.05	.04	4.505	6
25	P. xvii. 94	17 19.6	73 35	5.84	.05	4.519	6
26	P. xvii. 95	17 19.6	74 18	5.91	0.04	4.522	6
27	P. xvii. 99	17 20.8	95 0	4.68	.07	2.475	5...4
28	B.A.C. 5894	17 21.0	82 18	6.09	.02	4.519	6
29	σ	17 21.1	85 46	4.45	.07	2.483	5
30	P. xvii. 112	17 23.2	89 35	5.28	.08	4.505	5...6
31	B.A.C. 5910	17 24.7	90 58	5.47	0.02	4.522	6
32	P. xvii. 127	17 25.8	87 12	5.75	.06	4.519	6
33	L. 31952	17 27.6	95 40	6.07	.06	4.522	6
34	L. 32015	17 28.7	73 35	5.62	.08	4.522	6
35	53	17 29.4	80 20	6.11	.06	4.508	6

No. 11. Σ 2096, Dist. 22″. The bright star observed.
No. 12. OΣ 315, Dist. 1″. Observed as one mass. Spectrum I a ! (Vogel.)
No. 14. 4.16 is the mean of two determinations (4.12, 4.19) made on 1882.475 with twenty extinctions, and on 1885.363 with six. H.P.=4.41.
No. 15. Birm., No. 395. The red colour is not salient. Spectrum II a ! ! ! (Vogel.) Suspected of variability. (See Ast. Nach., xciv. p. 248.)
No. 16. Birm., No. 399. The star is reddish. Variable? (See Uranometria Argentina, p. 306.)
No. 18. The magnitude, after applying the mean absorption correction, is 2.42.
No. 19. Spectrum III a ! (Vogel.)
No. 20. Var. (U) Ophiuchi. Variability discovered by Sawyer in 1881. Period 20ʰ 8ᵐ.
Nos. 21, 22, and 24. Spectrum II a ! (Vogel.)
No. 23. The magnitude, after applying the mean absorption correction, is 2.83.
Nos. 25 and 30. Spectrum I a ! (Vogel.)
No. 26. Σ 2160, Dist. 4″. Observed as one mass.
No. 29. Spectrum II a !! (Vogel.)
No. 31. Σ 2173, Dist. <1″. Binary. Period 45 years. Observed as one mass.
No. 35. Σ 34¹, Dist. 41″. The brighter star observed. Spectrum I a ! (Vogel.)

Reference Number.	Star's Designation.	R.A. 1890.	N.P.D. 1890.	Adopted Zenithal Magnitude. Polaris 2.05.	Average Deviation in Magnitude.	Date, 1880 +	Mag. Argel. Uran.
		h. m.	° ′				
36	α	17 29.8	77 22	2.23	0.07	2.483	2
37	μ	17 31.9	98 3	4.71	.06	2.483	5...4
38	P. xvii. 203	17 37.0	74 0	5.74	.10	4.519	6
39	β	17 38.0	85 23	2.92	.06	...	3
40	61 1st star	17 39.0	87 22	6.34	.04	4.522	} 6
41	61 2nd star	17 39.0	87 22	6.54	0.03	4.522	
42	L. 32408	17 39.0	75 32	6.14	.04	4.522	6
43	γ	17 42.4	87 15	3.83	.08	2.483	4 ..3
44	ν	17 53.0	99 46	3.48	.05	2.483	4...3
45	66	17 54.8	85 37	4.75	.10	4.519	5
46	67	17 55.1	87 4	4.35	0.03	...	4
47	68	17 56.2	88 41	4.75	.05	2.483	4...5
48	τ	17 57.1	98 11	5.08	.04	4.519	5
49	70	17 59.9	87 28	4.35	.02	...	4...5
50	71	18 2.1	81 17	4.91	.03	4.508	5
51	72	18 2.1	80 27	3.92	0.04	2.483	3...4
52	73	18 4.1	86 2	5.82	.04	4.502	6
53	W.B. xviii. 28	18 4.4	86 53	5.86	.03	4.522	6
54	B.A.C. 6213	18 13.8	82 49	5.96	.05	4.502	6
55	74	18 15.4	86 41	5.06	.07	4.519	5
56	L. 34021	18 20.4	82 1	6.10	0.05	4.522	6
57	W.B. xviii. 10	18 31.0	80 57	5.60	.04	4.502	6
58	L. 34486	18 31.0	83 25	5.39	.08	4.502	6

No. 36. Spectrum I a !! (Vogel.)
No. 39. Birm., No. 417. The red colour is not salient. 2.92 is the mean of three accordant determinations made on 1882.483, 1882.732, and 1882.735. Spectrum II a !!! (Vogel.)
Nos. 40 and 41. These stars form Σ 2202, Dist. 20″.5.
No. 42. A star of about the same brilliancy precedes nearly one minute in R.A. and of less declination, which is not given in the Uranometria Nova. Spectrum I a ! (Vogel.)
No. 43. Spectrum I a ! (Vogel.)
No. 46. Σ Unum., Dist. 55″. The brighter star observed. 4.35 is the mean of two determinations (4.35, 4.36) made on 1882.483 with twenty extinctions, and on 1885.363 with six. H.P.=4.02.
No. 47. Variable? (See Uran. Argent., p. 307.) A determination made on 1885.363 with six extinctions gave the magnitude 4.69. H.P.=4.42. Spectrum I a ! (Vogel.)
No. 48. Σ 2262, Dist. 1″.5. Binary. Period 218 years. Observed as one mass.
No. 49. Σ 2272, Dist. 3″. Binary. Period 94 years. Observed as one mass. 4.35 is the mean of two determinations (4.40, 4 30) made on 1882.483 with twenty extinctions, and on 1885.363 with six. H.P.=4.11. Spectrum II a !! (Vogel.)
No. 51. Spectrum I a ! (Vogel.)
No. 52. Σ 2281, Dist. 1″.7. Observed as one mass. Spectrum I a ! (Vogel.)
Nos. 53 and 55. Spectrum II a ! (Vogel.)

ORION.

Reference Number.	Star's Designation.	R.A. 1890.	N.P.D. 1890.	Adopted Zenithal Magnitude. Polaris 2.05.	Average Deviation in Magnitude.	Date. 1880 +	Mag. Argel. Uran.
		h. m.	° ′				
1	P. iv. 169	4 38.3	79 3	5.56	0.05	4.177	6
2	L. 8943	4 39.9	78 29	5.65	.07	4.177	6
3	L. 9037	4 43.0	86 36	6.16	.06	4.064	6
4	π^3	4 43.9	83 14	3.62	.09	3.007	4
5	π^2	4 44.6	81 17	4.86	.09	...	5...4
6	π^4	4 45.3	84 35	3.99	0.17	1.976	4...5
7	P. iv. 214	4 45.7	80 12	6.07	.03	4.064	6
8	o^1	4 46.3	75 56	5.40	.09	4.177	5...6
9	π^5	4 48.5	87 44	4.10	.05	1.976	4
10	6....................	4 48.7	78 45	5.04	.13	4.177	6
11	π^1	4 48.8	80 1	5.04	0.05	4.064	5
12	P. iv. 236	4 48.9	82 23	5.85	.06	5.206	6
13	P. iv. 239	4 49.2	89 43	5.99	.14	3.987	6
14	o^2	4 50.2	76 39	4.48	.04	4.177	5
15	π^6	4 52.9	88 27	4.74	.07	1.976	5...4
16	L. 9419 1st star...	4 54.7	86 33	6.00	0.07	4.064	6
17	L. 9419 2nd star	4 54.8	86 33	6.60	.10	3.987	5
18	P. iv. 276	4 56.2	89 25	6.29	.08	4.064	6
19	11	4 58.3	74 45	4.76	.04	4.177	5
20	L. 9579	4 59.4	93 12	6.00	.07	4.064	6

No. 1.　B.A.C. assigns this star to Taurus.　Spectrum I a! (Vogel.)

No. 4.　Variable ?　(See Uran. Argent., p. 325.)　A determination made on 1885.216 with six extinctions, gave the magnitude 3.60.　H.P. = 3.33.

No. 5.　4.86 is the mean of two determinations (4.84, 4.89) made on 1881.976 with twenty extinctions, and on 1885.216 with six.　H.P. = 4.42.　Spectrum I a! (Vogel.)

No. 6.　Variable ?　Gilliss.　(See Astron. Obs., p. 663.)

No. 8.　Birm., No. 87.　The colour is a decided orange.　Spectrum III a !!! (Vogel.)

Nos. 10, 11, and 19.　Spectrum I a! (Vogel.)

No. 12.　Birm., No. 89.　The red colour is not salient.

No. 14.　Spectrum II a ! (Vogel.)

Nos. 16 and 17.　These stars form Σ 627, Dist. 21″.

No. 18　Birm., No. 95.　The colour is decidedly orange.

Reference Number.	Star's Designation.	R.A. 1890.	N.P.D. 1890.	Adopted Zenithal Magnitude. Polaris 2.05.	Average Deviation in Magnitude.	Date, 1880 +	Mag. Argel. Uran.
		h. m.	° ′				
21	14	5 1.9	81 39	5.48	0.04	4.064	6
22	16	5 3.3	80 19	5.43	.06	4.064	6
23	15	5 3.4	74 33	4.78	.05	4.177	5...6
24	L. 9744	5 4.5	90 43	6.43	.04	4.125	6
25	P. v. 1	5 5.4	74 5	5.31	.05	4.177	6
26	L. 9764	5 5.8	92 38	6.06	0.02	4.125	6
27	W.B. v. 64........	5 6.1	89 6	6.03	.03	4.125	6
28	ρ................	5 7.5	87 16	4.40	.02	4.152	5
29	L. 9820	5 8.9	84 58	6.02	.05	4.125	6
30	β	5 9.3	98 20	+1.03	.07	...	1
31	W.B. v. 169	5 9.3	91 32	6.21	0.03	4.141	6
32	18	5 9.8	78 47	5.80	.03	4.177	6
33	τ................	5 12.3	96 58	3.95	.10	...	4
34	21	5 13.3	87 31	5.33	.04	4.125	6...5
35	L. 9973	5 14.2	91 32	6.32	.05	4.141	6
36	B.A.C. 1656	5 15.6	91 41	6.00	0.04	4.152	6
37	ο................	5 16.2	90 29	4.77	.05	4.141	5
38	23	5 17.1	86 34	5.20	.09	5.206	5...6
39	29	5 18.7	97 54	4.18	.08	4.155	5...4
40	27	5 18.9	91 0	5.25	.04	4.177	6
41	η	5 19.0	92 30	3.66	0.09	2.073	3...4
42	ψ¹	5 19.0	88 15	4.61	.03	4.152	5
43	γ	5 19.2	83 45	1.79	.06	2.073	2
44	L. 10184	5 19.9	100 26	5.89	.09	4.155	6
45	ψ²	5 21.1	87 0	4.63	.04	4.141	5

No. 21. OΣ 98, Dist. 1″. Binary. Observed as one mass. Spectrum I a ! (Vogel.)

No. 22. Spectrum I a ! (Vogel.)

No. 24. Birm., No. 100. The red colour is not salient.

No. 27. Σ 652, Dist. 1″.7. Observed as one mass.

No. 30. Σ 668, Dist. 9″. Observed as one mass. Seidel suggests variability. (See Result. Phot. Mess., p. 160.) +1.03 is the mean of seven determinations, made on as many nights, four at Oxford and three at Cairo, involving eighty extinctions between 1881.976 and 1883.127. H.P. = +0.68. For explanation of notation see Preface.

No. 32. Spectrum I a ! (Vogel.)

No. 33. 3.95 is the mean of two determinations (3.99, 3.90) made on 1882.043 with twenty extinctions, and on 1885.189 with six. H.P. = 3.65.

No. 34. Variable? (See Uran. Argent., p. 326.)

No. 38. Σ 696, Dist. 32″. The brighter star observed.

No. 41. Close double. Observed as one mass.

No. 43. Variable? (See Uran. Argent., p. 327.)

Reference Number.	Star's Designation.	R.A. 1890.	N.P.D. 1890.	Adopted Zenithal Magnitude. Polaris 2.05.	Average Deviation in Magnitude.	Date, 1880 +	Mag. Argel. Uran.
		h. m.	o ′				
46	L. 10264	5 22.4	88 47	6.27	0.08	4.152	6
47	31	5 24.2	91 11	5.02	.04	4.155	5
48	L. 10328	5 24.2	88 18	5.59	.04	4.141	6
49	32	5 24.9	84 8	4.36	.07	4.152	5...6
50	33	5 25.5	86 47	5.05	.06	4.155	6
51	δ	5 26.4	90 23	2.02	0.06	2.073	2
52	v	5 26.6	97 23	5.18	.03	2.221	5...4
53	L. 10437	5 27.2	91 40	5.24	.03	4.155	6
54	35	5 27.7	75 46	5.64	.07	4.177	6
55	38	5 28.5	86 18	5.39	.04	4.064	6
56	ϕ^1	5 28.8	80 35	4.55	0.09	5.137	5
57	λ	5 29.1	80 8	3.52	.08	3.040	3...4
58	θ^1	5 29.9	95 28	4.61	.06	4.122	} 4
59	θ^2	5 30.0	95 30	5.14	.10	4.122	
60	42	5 30.0	94 55	5.45	.08	2.043	5...4
61	ι	5 30.1	95 59	3.19	0.06	2.073	3
62	45	5 30.2	94 56	5.78	.07	2.221	5...4
63	ϵ	5 30.6	91 16	1.85	.04	2.221	2
64	ϕ^2	5 30.9	80 46	4.66	.07	2.043	5...4
65	L. 10600	5 32.1	82 31	5.84	.03	5.206	6

No. 47. Σ 725, Dist. 13″. Observed as one mass. Birm., No. 110. The red colour is not salient. Variable? (See Uran. Argent., p. 328.)

No. 49. Σ 728, Dist. 1″. Observed as one mass.

No. 51. Σ 14¹, Dist. 53″. The brighter star observed. Variability discovery by Herschel in 1843. Max. 2.2, Min. 2.7. Period irregular. A determination made on 1885.189 gave the magnitude 2.29. H.P.=2.36. Spectrum I b ! (Vogel).

No. 52. A determination made on 1885.137 gave the magnitude 4.91. H.P.=4.66.

No. 54. Spectrum I a ! (Vogel.)

No. 56. The determination of 1882.221 (see Memoirs R.A.S., vol. xlvii, p. 443) is rejected. Clouds were passing at the time of the observation.

No. 57. Σ 738, Dist. 4″. Observed as one mass. Variable? (See Uranometria Argentina, p. 328.)

Nos. 60 and 62. Probably variable. (See Uran. Argent., p. 329.) H.P.=4.60 and 4.95 respectively.

No. 61. Σ 752, Dist. 11″. Observed as one mass.

No. 63. 1.85 is the mean of two determinations (1.87, 1.84`, one at Oxford involving twenty extinctions and one at Cairo involving ten.

No. 64. Birm., No. 113. The red colour is not salient. 4.66 is the mean of two determinations (4.70, 4.62) made on 1882.043 with twenty extinctions, and on 1883.040 with ten.

Reference Number.	Star's Designation.	R.A. 1890.	N.P.D. 1890.	Adopted Zenithal Magnitude. Polaris 2.05.	Average Deviation in Magnitude.	Date, 1880+	Mag. Argel. Uran.
		h. m.	° ′				
66	L. 10622	5 32.1	101 51	6.07	0.05	4.155	6
67	σ	5 33.2	92 40	3.88	.03	5.137	4...3
68	ω	5 33.4	85 56	4.33	.04	4.122	5
69	49	5 33.6	97 16	5.18	.04	4.064	5
70	P. v. 178	5 34.0	93 37	6.07	.04	4.179	6
71	ζ..................	5 35.2	92 0	1.80	0.03	2.221	2
72	L. 10734	5 35.3	91 11	5.27	.06	4.179	6
73	51	5 36.8	88 35	5.49	.02	4.155	6...5
74	L. 10826	5 37.6	96 51	6.04	.03	4.125	6
75	P. v. 206	5 39.2	86 2	6.09	.04	4.079	6
76	B.A.C. 1826	5 40.8	80 31	6.02	0.07	4.177	6
77	P. v. 220	5 40.9	88 52	6.12	.06	4.179	6
78	52	5 42.1	83 36	5.18	.03	4.125	6...5
79	κ	5 42.5	99 42	2.42	.06	2.221	3...2
80	W.B. v. 1048......	5 43.1	94 7	6.00	.09	...	6
81	P. v. 239	5 44.0	80 10	5.67	0.05	4.177	6
82	L. 11061	5 44.4	85 37	6.00	.06	4.155	6
83	55	5 46.1	97 33	5.32	.04	4.097	6...5
84	56	5 46.7	88 10	5.08	.09	4.097	6...5
85	χ¹	5 47.9	69 45	5.06	.09	....	5...4
86	57	5 48.4	70 16	6.03	0.04	4.179	6
87	α	5 49.2	82 37	+ 0.02	.05	...	1
88	L. 11221	5 50.1	94 38	5.82	.07	4.097	6
89	B.A.C. 1893	5 50.4	80 30	6.08	.09	4.097	6
90	L. 11224	5 50.7	78 30	6.22	.05	4.179	6

No. 67. Σ 762. Triple. The closer pair, dist. 12″, observed as one mass. Birm., No. 116. The red colour is not salient.

No. 71. Σ 774. Triple. The closer pair, dist. 2″.5, observed as one mass.

No. 73. Birm., No. 119. The colour is slightly red.

No. 75. Spectrum I a ! (Vogel.)

No. 78. Σ 795, Dist. 1″.7. Observed as one mass.

No. 80. 6.00 is the mean of two determinations (6.03, 5.97) made on 1884.179 with twenty extinctions, and on 1885.189 with six. H.P. = 5.68.

Nos. 82 and 84. Called red in Uranometria Argentina.

No. 85. 5.06 is the mean of two determinations (5.10, 4.99) made on 1882.221 with twenty extinctions, and on 1885.206 with six. H.P. = 4.65.

No. 87. Variability discovered by Herschel in 1836. Max. 1.0, Min. 1.4 mag. Period irregular. Birm., No. 127. The colour is reddish. Spectrum III a ! ! ! (Vogel.) + 0.02 is the mean of seven determinations, made on as many nights, four at Oxford and three at Cairo, involving eighty extinctions made between 1882.221 and 1883.169. H.P. = + 0.00. For explanation of notation see Preface.

Refer- ence Number.	Star's Designation.	R.A. 1890.	N.P.D. 1890.	Adopted Zenithal Magnitude. Polaris 2.05.	Average Deviation in Magnitude.	Date, 1880 +	Mag. Argel. Uran.
		h. m.'	o '				
91	B.A.C. 1907	5 52.7	77 12	5.63	0.06	4.155	6
92	59	5 52.7	88 10	6.07	.04	4.097	6
93	60	5 53.2	89 27	5.13	.05	4.125	6...5
94	L. 11382	5 54.5	93 5	5.04	.04	4.097	5...6
95	μ	5 56.3	80 21	4.69	.10	...	5...4
96	64	5 57.0	70 18	5.00	0.08	4.179	6
97	L. 11455	5 57.3	78 19	6.06	.02	4.212	6
98	χ²	5 57.4	69 52	4.88	.07	4.212	5
99	L. 11530	5 58.9	96 42	5.29	.03	4.097	6
100	63	5 59.1	84 34	5.65	.05	4.125	6
101	66	5 59.2	85 50	5.72	0.07	4.097	6
102	L. 11621	6 1.2	94 11	5.23	.08	4.125	6
103	ν.................	6 1.3	75 13	4.45	.04	5.137	5...4
104	L. 11688	6 3.3	87 29	5.68	.11	4.125	.6
105	L. 11748	6 5.3	76 20	5.61	.06	4.212	6
106	ξ.................	6 5.5	75 46	4.41	0.08	5.137	5...4
107	68	6 5.5	70 11	5.67	.05	4.179	6
108	69	6 5.6	73 51	5.24	.03	4.212	6...5
109	71	6 8.4	70 48	5.28	.05	4.218	6...5
110	L. 11884	6 8.9	76 7	5.69	.06	4.097	6
111	72	6 9.1	73 49	5.38	0.10	4.212	6...5
112	73	6 9.6	77 25	5.33	.07	4.218	6
113	L. 11936	6 10.0	90 28	5.69	.07	4.125	6
114	L. 11947	6 10.2	88 48	6.34	.03	4.141	6
115	74	6 10.3	77 42	5.27	.08	4.179	5...6
116	75	6 11.1	80 1	5.27	0.04	4.125	6
117	L. 12057	6 13.8	75 18	6.08	.04	4.212	6
118	77	6 21.6	89 38	5.04	.07	4.141	6

No. 93. Spectrum I a ! (Vogel.)
No. 94. Variable ! (See Uran. Argent., p. 330.) H.P. = 4.73.
No. 95. 4.69 is the mean of two determinations (4.83, 4.55) made on 1882.048 with twenty extinctions, and on 1885.206 with six. H.P. = 4.29. Spectrum I a ! (Vogel.)
No. 103. The determination of 1882.043 (see Memoirs R.A.S., vol. xlvii, p. 444) is rejected. The night was uncertain.
No. 104. Σ 855, Dist. 29". Observed as one mass. Spectrum I a ! (Vogel.)
No. 106. The determination of 1882.221 (see Memoirs R.A.S., vol. xlvii, p. 444) is rejected. The night was very doubtful. Spectrum I a ! (Vogel.)
Nos. 107 and 116. Spectrum I a ! (Vogel.)
No. 117. Birm., No. 142. The red colour is not salient. Spectrum III a ! (Vogel.)

PEGASUS.

Reference Number.	Star's Designation.	R.A. 1890.	N.P.D. 1890.	Adopted Zenithal Magnitude. Polaris 2.05.	Average Deviation in Magnitude.	Date, 1880 +	Mag. Argel. Uran.
		h. m.	° ′				
1	W.B. xxi. 319 ...	21 15.3	68 26	5.77	0.05	3.656	6
2	B.A.C. 7410	21 16.1	66 36	5.90	.04	3.656	6
3	1	21 17.0	70 40	4.30	.08	2.645	4...5
4	B.A.C. 7437	21 19.0	66 12	5.61	.05	3.656	6
5	2	21 25.0	66 51	4.52	.02	...	5
6	W.B. xxi. 557 ...	21 25.8	78 20	5.97	0.05	3.645	6
7	3	21 32.2	83 53	6.03	.07	3.639	6
8	4	21 33.0	84 44	5.59	.03	5.549	6...5
9	7	21 36.8	84 49	5.41	.05	3.639	6...5
10	Bradley 2827......	21 37.2	79 41	5.93	.09	3.645	6
11	ε...................	21 38.8	80 38	2.43	0.04	...	2...3
12	9	21 39.3	73 9	4.44	.08	2.656	5
13	κ	21 39.7	64 52	4.16	.02	2.656	4
14	12	21 41.0	67 33	5.44	.04	3.661	6 . 5
15	11	21 41.7	87 49	5.55	.02	3.639	6...5
16	13	21 44.9	73 13	5.56	0.03	3.645	6
17	14	21 45.0	60 20	5.16	.12	3.661	5
18	W.B. xxi. 1096...	21 46.4	70 42	5.73	.08	3.639	6
19	15	21 47.6	61 43	5.92	.06	...	6...5
20	16	21 48.1	64 36	5.18	.06	3.661	5...6

Nos. 2 and 4. B.A.C. assigns these stars to Vulpecula.

No. 3. Σ 11^{II}, Dist. 36″. The brighter star observed.

No. 5. Birm., No. 583. The colour is slightly orange. 4.52 is the mean of two determinations (4.48, 4.57) made on 1882.645 with twenty extinctions, and on 1885.549 with six.

No. 7. Σ 56^{I}, Dist. 39″. The brighter star observed.

No. 8. Spectrum I a ! (Vogel.)

No. 9. Spectrum III a !! (Vogel.)

No. 11. Triple. The bright star only observed. Birm., No. 591, and also called red in Uran. Argent. The red colour however is not salient. Variable? See Ast. Nach., vol. lxi, p. 136, and Uran. Argent., p. 338. Spectrum II a ! ! ! (Vogel.) 2.43 is the mean of four accordant determinations made between 1882.645 and 1882.661, involving eighty extinctions.

No. 13. Σ 2824, Dist. 12″. The brighter star observed.

No. 15. 11 Pegasi = 27 Aquarii. See Introd. to B.A.C., p. 75.

No. 19. 5.92 is the mean of two determinations (5.98, 5.86) made on 1883.661 with twenty extinctions, and on 1885.565 with six. H.P. = 5.63.

Reference Number.	Star's Designation.	R.A. 1890.	N.P.D. 1890.	Adopted Zenithal Magnitude. Polaris 2.05.	Average Deviation in Magnitude.	Date, 1880 +	Mag. Argel. Uran.
		h. m.	° ′				
21	W.B. xxi. 1136 ...	21 48.4	70 51	5.74	0.05	3.661	6
22	17	21 51.6	78 27	5.36	.05	3.735	6...5
23	18	21 54.6	83 49	6.04	.04	3.639	6
24	19	21 55.7	82 16	5.93	.03	3.639	6
25	20	21 55.7	77 24	5.68	.07	3.735	6...5
26	21	21 57.9	79 9	5.72	0.04	5.565	6...5
27	ν	22 0.1	85 29	4.84	.12	3 639	5
28	L. 43081	22 0.2	63 51	5.95	.02	3.735	6
29	23	22 0.6	61 34	5.31	.03	3.735	6...5
30	ι	22 1.9	65 12	4.24	.08	...	4
31	θ	22 4.6	84 21	3.53	0.03	2.658	3...4
32	π	22 5.1	57 22	4.06	.03	...	4
33	P. xxii. 29	22 7.9	55 56	5.23	.08	3.754	6...5
34	P. xxii. 32	22 8.6	61 56	6.15	.05	3.754	6
35	30	22 14.9	84 46	4.99	.09	3.757	5...6
36	31	22 16.1	78 21	5.09	0.11	5.565	5...4
37	32	22 16.3	62 13	5.07	.06	3.735	5
38	34	22 21.0	86 10	5.90	.04	3.757	6
39	35	22 22.3	85 52	5.02	.04	3.771	6...5
40	P. xxii. 113	22 22.7	58 43	6.19	.04	3.779	6
41	36	22 23.7	81 26	5.89	0.05	3.771	6
42	P. xxii. 120	22 24.0	63 48	5.85	.07	3.735	6
43	37	22 24.4	86 8	5.05	.04	3.771	6...5
44	38	22 25.0	57 59	5.62	.06	5.565	6...5
45	39	22 27.3	70 20	6.24	.04	3.735	6
46	40	22 33.6	71 3	5.95	0.06	3.735	6
47	41	22 34.5	70 53	6.00	.10	3.735	6
48	P. xxii. 186	22 35.4	76 1	5.67	.02	3.757	6...5
49	ζ	22 36.0	79 45	3.29	.01	...	3...4
50	P. xxii. 195	22 36.5	76 3	6.02	.05	3.757	6

No. 24. Called red in Uranometria Argentina.
No. 27. Spectrum III a ! (Vogel.)
No. 30. 4.24 is the mean of two determinations (4.25, 4.22) made on 1882.656 with twenty extinctions, and on 1885.565 with six. H.P. = 3.99.
No. 31. Variable? See Uran. Argent., p. 338. Spectrum I a ! ! (Vogel.)
No. 32. 4.06 is the mean of two determinations (4.09, 4.04) made on 1882.658 with twenty extinctions, and on 1885.549 with six. H.P. = 4.41.
Nos. 39 and 41. These stars are called red in Uranometria Argentina.
No. 43. Σ 2912, Dist. 1″.1. Observed as one mass.
No. 49. 3.29 is the mean of two determinations (3.30, 3.27) made on 1882.658 with twenty extinctions, and on 1885.549 with six. Spectrum I a ! (Vogel.)

Reference Number.	Star's Designation.	R.A. 1890.	N.P.D. 1890.	Adopted Zenithal Magnitude. Polaris 2.05.	Average Deviation in Magnitude.	Date, 1880 +	Mag. Argel. Uran.
		h. m.	o ,				
51	o	22 36.6	61 16	5.15	0.05	3.735	5
52	η	22 37.9	60 21	2.95	.05	2.658	3
53	45	22 40.1	71 13	6.32	.04	3.754	6
54	ξ	22 41.2	78 23	4.16	.05	5.565	5...4
55	λ	22 41.2	67 1	4.05	.07	2.661	4
56	μ	22 44.7	65 59	3.58	0.02	2.661	4
57	σ	22 46.8	80 45	5.12	.03	3.771	5
58	P. xxii. 241	22 47.6	73 45	5.94	.06	3.754	6
59	ρ	22 49.7	81 46	5.02	.05	3.771	5
60	51	22 52.1	69 49	5.56	.02	3.754	6...5
61	52	22 53.7	78 52	6.22	0.01	...	6
62	P. xxii. 285	22 57.4	67 28	7.25	.03	3.793	6
63	β	22 58.4	62 31	2.50	.07	2.661	2...3
64	α	22 59.3	75 23	2.33	.08	...	2
65	55	23 1.5	81 11	4.72	.03	3.771	5
66	56	23 1.8	65 8	4.69	0.11	3.793	5
67	W.B. xxii. 1378	23 2.1	69 27	6.05	.01	3.793	6
68	57	23 4.0	81 55	5.33	.05	3.771	5...6
69	58	23 4.5	80 46	5.21	.02	3.771	5...6
70	P. xxiii. 4	23 5.3	73 0	6.03	.04	3.754	6
71	59	23 6.2	81 53	5.31	0.07	3.771	5
72	60	23 6.5	63 45	6.28	.06	3.793	6
73	W.B. xxiii. 137	23 9.2	66 30	6.21	.05	3.793	6
74	W.B. xxiii. 169	23 10.6	65 49	6.39	.02	3.793	6
75	τ	23 15.2	66 52	4.75	.07	2.661	5...4

No. 54. The determination of 1882.658 is rejected as erroneous. See Memoirs R.A.S., vol. xlvii, p. 445.
No. 59. Spectrum I a ! (Vogel.)
No. 61. OΣ 483, Dist. 1".2. Observed as one mass. 6.22 is the mean of two determinations (6.22, 6.21) made on 1883.757 with twenty extinctions, and on 1885.655 with six. H.P. = 5.86.
No. 63. Variable. Discovered by Schmidt in 1847. Max. 2.2, Min. 2.7 mag. Period irregular. Birm., No. 627. The red colour is not salient.
No. 64. 2.33 is the mean of two determinations (2.24, 2.42) made on 1882.661 with twenty extinctions, and on 1885.568 with six. H.P. = 2.61. Spectrum I a !! (Vogel.)
No. 65. Birm., No. 629. The colour is fine orange. Spectrum III a ! (Vogel.)
No. 68. Σ 2982, Dist. 35". The brighter star observed. Spectrum III a !! (Vogel.)
No. 70. Spectrum II a ! (Vogel.)
No. 71. Spectrum I a ! (Vogel.)

Reference Number.	Star's Designation.	R.A. 1890.	N.P.D. 1890.	Adopted Zenithal Magnitude. Polaris 2.05.	Average Deviation in Magnitude.	Date, 1880+	Mag. Argel. Uran.
		h. m.	° ′				
76	63	23 15.4	60 11	5.82	0.02	3.793	6
77	64	23 16.5	58 47	5.57	.04	3.793	6
78	65	23 17.2	69 46	6.29	.09	3.754	6
79	66	23 17.5	78 17	5.42	.07	3.771	6...5
80	67	23 19.5	58 13	5.66	.11	3.793	6
81	υ	23 19.9	67 12	4.57	0.08	2.661	5...4
82	69	23 22.2	65 26	6.04	.02	3.793	6
83	70	23 23.6	77 51	5.04	.04	...	5
84	71	23 28.0	68 6	5.55	.09	3.754	6
85	72	23 28.5	59 17	5.44	.06	3.793	6
86	73	23 29.2	57 7	6.03	0.05	3.801	6
87	W.B. xxiii. 593...	23 29.5	66 11	6.45	.13	3.754	6
88	W.B. xxiii. 613...	23 30.4	66 3	6.46	.04	3.801	6
89	W.B. xxiii. 629...	23 31.0	57 42	6.49	.10	3.782	6
90	74	23 32.1	73 47	6.14	.15	3.754	6
91	75	23 32.4	72 12	5.48	0.04	3.754	6...5
92	P. xxiii. 146	23 34.3	80 56	6.17	.03	3.779	6
93	77	23 37.8	80 17	5.26	.05	3.779	6...5
94	78	23 38.5	61 15	5.03	.09	3.801	5
95	79	23 44.1	61 46	6.04	.02	3.782	6
96	Bradley 3175......	23 46.8	68 56	6.05	0.04	3.801	6
97	φ	23 46.9	71 29	5.45	.06	5.589	6...5
98	82	23 47.0	79 40	5.42	.05	3.779	6...5
99	W.B. xxiii. 928...	23 47.4	72 43	6.58	.03	3.802	6
100	P. xxiii. 235	23 51.1	67 58	5.98	.12	3.782	6
101	ψ	23 52.2	65 28	4.70	0.04	2.661	5
102	85	23 56.4	63 29	5.72	.02	3.782	6
103	86	0 0.1	77 13	5.55	.12	3.782	6
104	W.B. xxiii. 1389	0 3.2	65 9	5.99	.07	3.793	6
105	87	0 3.4	72 25	5.53	.09	3.771	6

No. 79. Spectrum II a ! (Vogel.)
No. 83. 5.04 is the mean of two determinations (5.06, 5.02) made on 1883.782 with twenty extinctions, and on 1885.589 with six. H.P.=4.05.
No. 84. Birm., No. 643. The star is very slightly red.
Nos. 90, 92, and 98. Spectrum I a ! (Vogel.) No. 91. Spectrum I a !! (Vogel.)
No. 93. Birm., No. 645. The colour is yellow. Spectrum III a ! (Vogel.)
No. 94. Birm., No. 647. The red colour is not salient.
No. 97. Spectrum III a !!! (Vogel.)
No. 100. Birm., No. 653. The colour is slightly red.
No. 101. Birm., No. 655. The red colour is not salient.
No. 102. Double. Dist. 14″. The brighter star observed.

Reference Number.	Star's Designation.	R.A. 1890.	N.P.D. 1890.	Adopted Zenithal Magnitude. Polaris 2.05.	Average Deviation in Magnitude.	Date, 1880 +	Mag. Argel. Uran.
		h. m.	o '				
106	γ	0 7.6	75 25	2.47	0.06	2.661	3...2
107	χ	0 8.9	70 24	4.87	.06	3.779	5

No. 106. Variable? (See Ast. Nach., xciii, p. 189 and Uran. Argent., p. 341.) Another determination on 1885.589 gave the magnitude 2.72. **H.P. = 3.04.**
No. 107. Spectrum III a ! (Vogel.)

PERSEUS.

1	υ	1 31.2	41 56	3.76	0.06	2.699	4...3
2	φ	1 36.8	39 52	4.29	.05	2.699	4
3	B.A.C. 547	1 42.6	42 39	5.85	.09	5.174	6
4	1	1 44.8	35 24	5.70	.07	5.188	6
5	2	1 45.1	39 45	5.68	.07	5.182	6
6	3	1 51.6	41 20	5.82	0.08	5.174	6
7	4	1 55.0	36 3	5.15	.10	5.188	5...6
8	6	2 6.3	39 27	5.51	.01	5.182	6
9	9	2 14.7	34 39	5.46	.04	5.182	6...5
10	12	2 35.3	50 16	5.02	.02	5.174	5
11	θ	2 36.6	41 14	4.26	0.04	2.699	4
12	14	2 36.9	46 10	5.78	.04	5.188	6
13	η	2 42.6	34 34	4.13	.04		4...3
14	16	2 43.6	52 8	4.77	.06	...	5...4
15	17	2 44.7	55 24	4.90	.05	5.174	5
16	τ	2 46.5	37 41	4.07	0.04	2.699	4
17	20	2 46.8	52 7	5.51	.03	5.182	6
18	21	2 50.6	58 30	5.38	.10	5.188	5
19	π	2 51.7	50 47	4.89	.04	5.174	5
20	24	2 52.3	55 15	5.28	.15	5.182	5...6

Nos. 1, 2, and 3. B.A.C. assigns these stars to Andromeda.
No. 3. Σ 162, Dist. 1".9 and 20". Triple. The closer pair observed as one mass.
No. 11. Σ 296, Dist. 15". Observed as one mass.
No. 13. Σ 307. Multiple. The bright star observed. Birm., No. 49. The red colour is not salient. 4.13 is the mean of ten determinations, made on as many nights, seven at Oxford and three at Cairo, involving one hundred and forty extinctions made between 1882.699 and 1883.207. **H.P. = 3.93.**
No. 14. 4.77 is the mean of two determinations (4.84, 4.70) made on 1882.699 with twenty extinctions, and on 1885.239 with six. **H.P. = 4.44.**
No. 17. Σ 318. Triple. Observed as one mass.

Reference Number.	Star's Designation.	R.A. 1890.	N.P.D. 1890.	Adopted Zenithal Magnitude. Polaris 2.05.	Average Deviation in Magnitude.	Date, 1880+	Mag. Argel. Uran.
		h. m.	° ′				
21	P. ii. 220	2 53.0	38 5	5.24	0.01	5.188	5
22	γ	2 56.8	36 56	3.06	.04	2.702	3
23	B.A.C. 948	2 57.3	33 44	4.98	.08	5.174	5
24	ρ.................	2 58.1	51 35	4.24	.05	2.702	4
25	ι.................	3 1.0	40 48	4.30	.08	2.702	4
26	β	3 1.0	49 28	2.40	0.06	2.702	Var.
27	κ	3 2.1	45 33	4.08	.05	2.702	4...5
28	ω	3 4.2	50 48	4.94	.03	5.224	5
29	D.M.+56°, No.798	3 7.4	33 17	5.86	.14	5.193	6
30	B.A.C. 995	3 8.3	39 28	5.25	.03	5.174	5
31	P. iii. 9	3 8.6	59 51	5.42	0.08	5.188	6
32	30	3 10.4	46 22	5.55	.10	5.182	6
33	29	3 10.8	40 11	5.40	.12	5.188	5
34	31	3 11.3	40 18	5.08	.09	5.188	
35	P. iii. 23	3 11.9	56 11	4.95	.04	5.224	5
36	32	3 14.1	47 4	5.00	0.03	5.182	5
37	α	3 16.5	40 32	1.93	.02	2.702	2
38	P. iii. 53	3 20.2	41 19	5.05	.08	5.224	6
39	34	3 21.5	40 52	5.00	.10	5.182	5
40	σ	3 22.8	42 23	4.73	.05	...	5
41	36	3 24.8	44 19	5.55	0.04	5.191	6
42	R. 998	3 25.1	45 31	5.90	.01	5.193	6
43	W.B. iii. 484	3 25.7	54 55	5.90	.10	5.182	6
44	ψ	3 28.7	42 10	4.48	.02	5.191	5
45	P. iii. 104	3 34.0	52 46	6.08	.01	5.188	6

No. 21. Σ 331, Dist. 12″. Observed as one mass.

No. 24. Variability discovered by Schmidt in 1854. Max. 3.4, Min. 4.2 mag. Period irregular. Birm., No. 53. The red colour is not salient. Another determination made on 1885.239 gave the magnitude 4.08. H.P. = 3.68.

No. 26. Variable. Max. 2.2, Min. 3.7 mag. Period 2ᵈ, 20ʰ 48ᵐ 53ˢ.67. Birm., No. 55. This star does not appear to be coloured.

No. 28. Birm., No. 56. The red colour is not salient.

No. 40. Birm., No. 62. The colour is deep yellow. 4.73 is the mean of two determinations (4.78, 4.68) made on 1882.702 with twenty extinctions, and on 1885.239 with six. H.P. = 4.39.

Reference Number.	Star's Designation.	R.A. 1890.	N.P.D. 1890.	Adopted Zenithal Magnitude. Polaris 2.05.	Average Deviation in Magnitude.	Date, 1880 +	Mag. Arcel. Uran.
		h. m.	° '				
46	δ.....................	3 35.1	42 34	3.11	0.04	2.656	3
47	40	3 35.4	56 23	5.04	.04	5.191	5
48	O.A. 4062	3 37.0	44 14	6.13	.07	5.188	6
49	D.M.+36°, No.742	3 37.4	53 53	5.66	.05	5.193	6
50	ο.....................	3 37.4	58 3	4.40	.08		4
51	ν......................	3 37.7	47 46	4.06	0.05	2.702	4
52	B.A.C. 1142	3 38.3	44 40	5.86	.02	5.191	6
53	R. 1071	3 41.6	46 23	5.89	.09	5.188	6
54	B.A.C. 1172	3 42.4	45 22	6.04	.01	5.191	6
55	42	3 42.6	57 15	5.41	.11	5.182	6...5
56	W.B. iii. 942......	3 44.9	55 58	5.93	0.03		6
57	ζ......................	3 47.2	58 27	3.09	.02	2.705	3
58	D.M.+51°.No.803	3 47.3	38 6	6.36	.13	5.193	...
59	P. iii. 186	3 48.0	42 27	5.45	.10	5.188	6...5
60	43	3 48.4	39 38	5.66	.07	5.191	5...6
61	L. 7206	3 49.4	55 14	5.65	0.05	5.193	6
62	ε......................	3 50.5	50 21	3.13	.05	...	3 4
63	ξ......................	3 51.8	54 31	4.31	.07	2.705	4
64	λ	3 58.4	39 56	4.39	.04	2.702	4...5
65	48	4 0.7	42 35	4.30	.07	2.702	4
66	50	4 1.3	52 16	5.59	0.10	5.188	6
67	W.B. iii. 1331	4 3.9	56 42	5.94	.10	5.191	6
68	μ	4 6.8	41 52	4.17	.03	2.702	4...5
69	52	4 7.4	49 49	4.85	.08	5.188	5
70	B.A.C. 1301	4 9.9	39 58	4.80	.12	5.193	5

No. 47. Σ 431, Dist. 20″. The brighter star observed.

No. 50. 4.40 is the mean of two determinations (4.47, 4.32) made on 1882.702 with twenty extinctions, and on 1885.239 with six. H.P.=4.01.

No. 56. 5.93 is the mean of two determinations (5.98, 5.88) made on 1885.193 and on 1885.224, each with six extinctions. H.P.=5.70.

No. 57. Σ 464, Dist. 12″.5. Observed as one mass.

No. 58. It is stated in the corrigenda to the Uranometria Nova that the star was not visible to the naked eye.

No. 62. Σ 471, Dist. 8″.4. Observed as one mass. 3.13 is the mean of three accordant determinations, made between 1882.705 and 1883.237, involving forty extinctions.

No. 68. OΣ 78, Dist. 15″. Observed as one mass.

Reference Number.	Star's Designation.	R.A. 1890.	N.P.D. 1890.	Adopted Zenithal Magnitude. Polaris 2.05.	Average Deviation in Magnitude.	Date. 1880 +	Mag. Argel. Uran.
		h. m.	o '				
71	B.A.C. 1314	4 11.9	39 21	5.54	0.08	5.237	6
72	54	4 13.3	55 42	5.31	.05	5.191	6
73	53 (*d*)	4 13.6	43 46	4.95	.11	5.188	5
74	55	4 17.3	56 7	5.47	.07	5.188	6
75	P. iv. 69............	4 19.1	58 49	5.04	.10	5.193	6
76	57 (*m*)	4 25.5	47 10	6.06	0.03	5.193	6
77	58	4 29.1	48 58	4.60	.02	5.237	5
78	R. 1283	4 33.2	41 55	5.66	.10	5.188	6
79	R. 1289	4 35.0	40 14	5.60	.04	5.188	6
80	59	4 35.1	46 51	5.51	.02	5.237	6

No. 75.   B.A.C. assigns this star to Taurus.
No. 77.   Birm., No. 80. The red colour is not salient.

PISCES.

1	2....................	22 53.8	89 38	5.60	0.10	3.877	6
2	3....................	22 55.0	90 24	6.48	.03	3.877	6
3	β	22 58.3	86 46	4.41	.05	2.604	5...4
4	5....................	23 3.0	88 28	5.45	.05	5.568	6
5	γ	23 11.4	87 19	3.63	.07	2.604	4
6	7....................	23 14.7	85 13	5.05	0.08	3.877	6
7	κ....................	23 21.3	89 21	5.17	.03	3.888	5...4
8	9....................	23 21.6	89 29	6.48	.06	3.888	...
9	θ	23 22.4	84 13	4.18	.05	2.604	4...5
10	14	23 28.5	91 51	6.13	.04	3.894	6
11	16	23 30.8	88 30	5.77	0.06	5.568	6
12	ι	23 34.3	84 58	4.25	.03	2.604	4...5
13	λ....................	23 36.4	88 49	4.94	.04	3.894	5
14	19	23 40.8	87 7	5.21	.06	3.888	6
15	20	23 42.3	93 22	5.80	.03	3.877	6

Nos. 7 and 13.   Spectrum I a ! (Vogel.)

No. 9.   Spectrum II a ! (Vogel.)

No. 14.   Birm., No. 648. The star is distinctly red. Variable? (See Uran. Argent., p. 334, and Dunsink Obs., iv, p. 57.) Spectrum III b !!! (Vogel.)

Refer-ence Number.	Star's Designation.	R.A. 1890.	N.P.D. 1890.	Adopted Zenithal Magnitude. Polaris 2.05.	Average Deviation in Magnitude.	Date, 1880+	Mag. Argel. Uran.
		h. m.	° ′				
16	21	23 43.8	89 32	6.11	0.07	3.888	6
17	25	23 47.4	88 31	6.05	.05	3.877	6
18	27	23 53.1	94 10	5.30	.02	...	5...6
19	ω	23 53.7	83 45	4.19	.06	2.604	4
20	L. 47041	23 54.2	90 53	6.68	.07	3.888	6
21	29	23 56.2	93 38	5.30	0.08	3.877	5...6
22	30	23 56.3	96 38	4.55	.07	2.607	5
23	32	23 56.9	82 8	5.72	.05	...	6
24	33	23 59 7	96 19	4.82	.09	3.888	5
25	34	0 4.4	79 28	5.55	.07	3.877	6
26	35	0 9.3	81 47	5.95	0.05	3.888	6
27	36	0 10.9	82 22	6.18	.08	3.894	6
28	41	0 14.9	82 25	5.44	.06	3.894	6...5
29	47	0 22.3	72 43	4.94	.02	...	6...5
30	48	0 22.5	74 10	6.47	.06	3.946	6
31	51	0 26.7	83 39	5.86	0.06	3.894	6
32	52	0 26.8	70 19	5.45	.10	3.946	6
33	53	0 31.1	75 22	6.00	.03	3 888	6
34	54	0 33.8	69 20	6.08	.04	3.946	
35	55	0 34.1	69 10	5.48	.05	3 946	6...5
36	57	0 40.8	75 7	5.73	0.05	3.894	6...5
37	58	0 41.3	78 37	6.00	.06	...	5
38	59	0 41.4	71 1	6.24	.06	3 946	6
39	P. o. 189	0 42.6	85 16	5.74	.12	3.888	6
40	δ	0 43.0	83 1	4.32	.06		4...5

Nos. 16 and 17. Spectrum I a ! (Vogel.)

No. 18. 5.30 is the mean of two determinations (5.34, 5.27) made on 1883.894 with twenty extinctions, and on 1885.568 with six. H.P. = 5.03.

No. 20. Spectrum III a ! (Vogel.)   No. 21. Gilliss suspects variability. (Ast. Obs., p. 671.)

Nos. 22 and 24. Called red in the Uranometria Argentina.

No. 23. 5.72 is the mean of two determinations (5.66, 5.79) made on 1883.894 with twenty extinctions, and on 1885.568 with six. Spectrum I a ! (Vogel.)

No. 25. Σ 5, Dist. 8″. Observed as one mass.

No. 26. Σ 12, Dist. 11″.5. Observed as one mass.

No. 29. 4.04 is the mean of two determinations (4.93, 4.95) made on 1883.946 with twenty extinctions, and on 1885.568 with six. H.P. = 5.37. Spectrum III a !!! (Vogel.)

No. 31. Σ 36, Dist. 27″.5. The brighter star observed. Spectrum I a ! (Vogel.)

No. 35. Σ 46, Dist. 6″. Observed as one mass.   No. 36. Spectrum III a !!! (Vogel.)

No. 37. 6.00 is the mean of two determinations (6.08, 5.92) made on 1883.888 with twenty extinctions, and on 1885.568 with six. H.P. = 5.70.

No. 38. Spectrum I a ! (Vogel.)

No. 40. 4.32 is the mean of two determinations (4.25, 4.40) made on 1882.624 with twenty extinctions, and on 1885.568 with six. H.P. = 4.58.

Reference Number.	Star's Designation.	R.A. 1890.	N.P.D. 1890.	Adopted Zenithal Magnitude. Polaris 2.05.	Average Deviation in Magnitude.	Date, 1880+	Mag. Argel. Uran.
		h. m.	° ′				
41	64	0 43.2	73 39	5.09	0.05	3.946	6...5
42	65	0 44.0	62 53	5.83	.10	3.954	6
43	66	0 48.8	71 24	6.07	.08	3.888	6
44	67	0 50.1	63 23	6.00	.03	3.960	6
45	68	0 51.9	61 36	5.58	.05	3.960	6
46	P. o. 243	0 52.1	76 54	6.29	0.05	3.888	6
47	σ	0 56.8	58 47	5.66	.03	3.960	5
48	ε...................	0 57.2	82 42	4.17	.04	...	4
49	72	0 59.3	75 39	5.84	.07	3.894	6
50	ψ^1 1st star	0 59.8	69 7	5.16	.06	2.609	} 5...4
51	ψ^1 2nd star	0 59.8	69 7	5.34	0.07	2.609	
52	77 1st star.........	1 0.1	85 41	6.53	.11	3.954	} 6
53	77 2nd star	1 0.2	85 41	6.02	.03	3.954	
54	75	1 0.8	77 38	6.19	.03	3.886	6
55	ψ^2	1 2.1	69 51	5.76	.05	3.888	6...5
56	80	1 2.7	84 56	5.60	0 0.3	3.954	6...5
57	υ^3	1 3.9	70 56	5.60	.06	3.888	6
58	82	1 5.1	59 10	4.93	.04	3.888	5
59	χ	1 5.5	69 33	4.87	.07	2.609	5...4
60	τ.....................	1 5.6	60 30	4.24	.07	...	4
61	φ	1 7.8	66 0	4.77	0.11	2.609	5
62	ζ	1 8.0	83 0	4.93	.03	2.609	5...4
63	87	1 8.3	74 27	5.72	.06	3.886	6
64	88	1 9.0	83 35	6.01	.05	3.954	6
65	89	1 12.1	86 58	5.34	.11	3.954	5...6

No. 41. Spectrum I a ! (Vogel.)

No. 42. Σ 61, Dist. 4″.5. Observed as one mass.

No. 43. OΣ 20, Dist. < 1″. Observed as one mass. Spectrum I a ! (Vogel.)

No. 47. 40 Andromedæ. (See Introduction to B.A.C., p. 75.)

No. 48. 4.17 is the mean of two determinations (4.07, 4.26) made on 1882.639 with twenty extinctions, and on 1885.568 with six. H.P. = 4.46.

Nos. 50 and 51. These stars form Σ 88, Dist. 30″.

Nos. 52 and 53. These stars form Σ 90, Dist. 33″.

Nos. 55, 56, and 65. Spectrum I a ! (Vogel.)

No. 60. 4.24 is the mean of two accordant determinations made on 1882.609, and on 1882.639 with twenty extinctions. H.P. = 4.73.

No. 61. Σ 99, Dist. 8″. Observed as one mass.

No. 62. Σ 100, Dist. 23″. Observed as one mass.

Reference Number.	Star's Designation.	R.A. 1890.	N.P.D. 1890.	Adopted Zenithal Magnitude. Polaris 2.05.	Average Deviation in Magnitude.	Date, 1880+	Mag. Argel. Uran.
		h. m.	o ′				
66	υ................	1 13.4	63 19	4.47	0.12	2.620	4
67	91	1 15.0	61 50	5.56	.06	3.877	5
68	W.B. i. 320	1 17.5	70 6	6.25	.08	3.886	6
69	ρ................	1 20.3	71 24	5.33	.05	3.886	5
70	94	1 20.8	71 20	5.46	.05	3.886	...
71	μ	1 24.4	84 25	5.43	0.04	3 954	5
72	η	1 25.6	75 13	3.71	.05	...	4...3
73	P. i. 120..........	1 30.0	73 8	6.09	.03	3.886	6
74	π	1 31.3	78 25	5.60	.04	3.877	6
75	105...............	1 33.7	74 9	5.95	.07	3.877	6
76	ν..	1 35.7	85 4	4.08	0 06	2.620	5...4
77	107...............	1 36.5	70 15	5.38	.06	3.877	5..6
78	ο	1 39.6	81 24	4.29	.03	2.620	4
79	ξ................	1 47.9	87 21	4 70	.03	2.623	4
80	α¹	1 56.4	87 46	3.71	.07		}
81	α²	1 56.4	87 46	4.70	0.03	2.623	} 3...4

No. 72. **3.71** is the mean of two accordant determinations made on 1882.620 and on 1882.639, each with twenty extinctions. Spectrum II a ! (Vogel.)

No. 76. = 51 Ceti. See Introduction to B.A.C., p. 75.) Birm., No. 29. The colour is yellow. Spectrum II a ! (Vogel.)

No. 77. = 2 Arietis. (See Introduction to B.A.C., p. 75.)

No. 80. **3.71** is the mean of two accordant determinations made on 1882.623 and on 1882.630, each with twenty extinctions.

Nos. 80 and 81. These stars form Σ 202, Dist. 3″.6. Spectrum I a ! (Vogel.)

PLEIADES.

Refer-ence Number.	Star's Designation.	R.A. 1890.	N.P.D. 1890.	Adopted Zenithal Magnitude. Polaris 2.05.	Average Deviation in Magnitude.	Date, 1880 +	Mag. Argel. Uran.
		h. m.	° ′				
1	Cœleno	3 38.3	66 3	5.34	0.03	3.908	...
2	Electra	3 38.3	66 14	3.96	.05	3.908	4
3	18 (*m*)	3 38.6	65 30	5.99	.04	3.954	...
4	Taygeta	3 38.7	65 53	4.54	.03	3.908	5
5	1....................	3 38.9	66 19	7.38	.01	3.921	...
6	4....................	3 39.1	66 1	7.44	0.03	3.951	
7	6....................	3 39.2	66 3	9.32	.11	3.951	...
8	Maia	3 39.3	65 59	3.90	.05	3.951	5
9	7....................	3 39.3	66 18	7.54	.03	3.951	...
10	Asterope *k*	3 39.4	65 47	5.98	.05	3 908	
11	Asterope *l*	3 39.5	65 49	6.46	0.05	3.908	
12	8....................	3 39.7	66 9	7.36	.05	3.944	
13	9....................	3 39.7	66 9	7.68	.02	3 944	
14	10	3 39.9	66 5	7.18	.06	3.921	
15	12	3 40.4	65 49	6.74	.08	3 927	
16	13	3 40.5	66 21	8.22	0.04	3.951	
17	15	3 40.7	66 13	8.09	.10	3.951	
18	17	3 40.8	66 37	6.84	.05	3.908	
19	18	3 40.8	66 12	7.01	.04	3.924	
20	24 (*p*)	3 40.8	66 14	6.04	.03	3.921	
21	19	3 40.8	66 32	6.78	0.04	3.924	
22	20	3 40.8	65 45	7.52	.04	3.908	
23	21	3 40.9	65 41	7.60	.08	3.951	...
24	22	3 40.9	66 26	6.80	.05	3.924	5
25	23	3 40.9	66 40	7.53	.05	3.951	...

The magnitudes of the stars in the Pleiades were compared directly with that of Merope, deter-mined by measurement to be 4.30 on the scale of Polaris = 2.05 mag.

Other valuations of these magnitudes made by other Astronomers will be found in Memoirs R.A.S., vol. xlviii, p. 272.

For the literature connected with the magnitude of the stars in the Pleiades, see Bessel, Beobachtungen Verschiedener Sterne der Plejaden.

Wolf, Annales de l'Observatoire de Paris, Mémoires, Tome xiv, deuxième partie.

Lindemann, Mémoires de l'Académie Impériale des Sciences de St Pétersbourg. Tome xxxii, No. 6.

Pritchard, Memoirs of the Royal Astronomical Society, vol. xlviii.

Pickering, Harvard Observations, vol. xiv, p. 398.

Reference Number.	Star's Designation.	R.A. 1890.	N.P.D. 1890.	Adopted Zenithal Magnitude. Polaris 2.05.	Average Deviation in Magnitude.	Date, 1880 +	Mag. Argel. Uran.
		h. m.	° ′				
26	24	3 41.0	66 3	6.53	0.03	3.924	...
27	η Tauri	3 41.0	66 14	3.12	.04	3.908	3
28	28	3 41.8	66 55	5.75	.03	3.908	...
29	29	3 42.0	66 0	6.58	.06	3.921	..
30	26 (s)	3 42.5	66 29	6.56	.06	3.927	
31	Atlas	3 42.6	66 17	4.00	0.06	3.951	4
32	Pleione	3 42.6	66 12	5.46	.07	3.908	...
33	30	3 42.7	66 27	7.89	.02	3.951	
34	31	3 42.7	65 56	6.81	.05	3.951	
35	32	3 42.8	65 57	6.34	.06	3.927	
36	33	3 42.9	66 5	6.78	0.04	3.951	
37	34	3 43.2	66 37	6.27	.08	3.927	
38	35	3 43.2	66 5	9.67	.04	3.954	
39	36	3 43.4	66 7	9.07	.13	3.954	
40	37	3 43.4	65 59	7.28	.01	3.927	
41	38	3 43.4	66 29	6.84	0.05	3.944	
42	39	3 43.9	65 50	7.33	.05	3.944	
43	40	3 44.3	66 22	7.17	.05	3.944	...

SAGITTA.

1	1	19 10.6	68 58	5.84	0.04	3.658	6
2	2	19 19.4	73 17	6.05	.06	3.658	6
3	3	19 19.8	73 15	6.31	.04	3.658	
4	ε	19 32.3	73 47	5.83	.06	3.655	6
5	5	19 35.2	72 14	4.42	.05	5.568	4..5
6	β	19 36.1	72 47	4.61	0.05	...	4..5
7	δ	19 42.5	71 44	3.83	.03	5.565	4
8	ζ	19 44.1	71 8	4.85	.02	3.655	5
9	10	19 51.0	73 39	5.29	.03	3.658	6
10	11	19 52.8	73 30	5.22	.04	3.594	6

No. 1. B.A.C. assigns this star to Vulpecula.
Nos. 2 and 3. These stars form Σ 41¹, Dist. 336″.
No. 4. Σ Unnum., Dist. 92″. The bright star observed. Spectrum II a ! (Vogel.)
Nos. 5, 7, and 11. The determinations of 1882.592 (see Memoirs of R.A.S., vol. xlvii, p. 448) are rejected. The meteorological conditions were variable.
No. 6. 4.61 is the mean of two determinations made on 1882.592 and on 1882.595, each with twenty extinctions. H.P. = 4.40.
No. 7. Spectrum III a !! (Vogel.)
No. 8. Σ 2585, Dist. 8″. Observed as one mass.

Reference Number.	Star's Designation.	R.A. 1890.	N.P.D. 1890.	Adopted Zenithal Magnitude. Polaris 2.05.	Average Deviation in Magnitude.	Date. 1880+	Mag. Argel. Uran.
		h. m.	° ′				
11	γ	19 53.9	70 48	3.72	0.06	5.565	4...3
12	13	19 55.1	72 47	5.35	.03	3.658	6
13	14	19 58.5	74 17	5.50	.11	3.655	6
14	15	19 59.2	73 13	6.00	.07	3.655	6
15	η	20 0.3	70 20	5.39	.09	3.592	5...6
16	θ	20 5.1	69 25	6.01	0.04	3.655	6
17	18	20 11.5	68 44	5.92	.08	3.658	6

No. 11. Spectrum II a ! (Vogel.)
No. 12. Double : dist. 29″. The brighter star observed. Spectrum III a ! (Vogel.)
No. 16. Σ 2637, Dist. 11″ and 70″. Triple. The closer pair observed as one mass.

SCORPIO.

1	a	16 22.7	116 11	1.13	0.07	...	1...2

No. 1. 1.13 is the mean of two determinations made at Cairo. See Memoirs R.A.S., vol. xlvii, p. 449.

SCUTUM.

1	W.B. xviii. 339 ...	18 17.7	98 59	5.13	0.06	4.601	6
2	B.A.C. 6325	18 29.2	98 19	4.18	.09	4.604	4...5
3	P. xviii. 149	18 36.2	99 9	4.96	.10	4.604	5
4	P. xviii. 157	18 37.5	98 23	5.20	.08	4.601	5
5	P. xviii. 177	18 41.3	94 52	4.53	.09	4.601	5...4
6	R. Scuti	18 41.6	95 49	5.37	0.06	4.604	Var.
7	B.A.C. 6464	18 51.2	95 59	5.15	.02	4.604	5

No. 2. B.A.C. assigns the star to Aquila. Birm., No. 449. The red colour is not salient.
Nos. 3, 4, 5, and 7. B.A.C. assigns these stars to Aquila.
No. 6. Variability discovered by Pigott in 1795. Max. 4.7 to 5.7, and Min. 6.0 to 8.5 mag. Period 71.1 days. Birm., No. 462. The colour is slightly red.
No. 7. Called red in Uranometria Argentina.

SERPENS.

Reference Number.	Star's Designation.	R.A. 1890.	N.P.D. 1890.	Adopted Zenithal Magnitude. Polaris 2.05.	Average Deviation in Magnitude.	Date, 1880+	Mag. Argel. Uran.
		h. m.	° ′				
1	3	15 9.7	84 39	5.55	0.05	4.491	6
2	4	15 10.2	89 13	5.59	.05	5.366	6
3	5	15 13.7	87 49	5.13	.05	4.491	5
4	6	15 15.4	88 53	5.63	.08	4.494	6
5	τ^1	15 20.7	74 11	5.41	.05	4.491	6
6	10	15 23.1	87 47	5.19	0.04	4.494	6
7	τ^2	15 27.1	73 34	6.05	.03	4.491	6
8	11 (Λ^1)	15 27.3	90 49	5.91	.04	4.489	6
9	δ	15 29.6	79 6	3.80	.05	2.467	3...4
10	W.B. xv. 505......	15 29.6	87 58	6.51	.09	4.500	6
11	τ^3	15 30.6	71 59	6.15	0.03	4.491	6
12	16	15 31.2	79 37	5.22	.07	4.500	6
13	τ^4	15 31.4	74 32	6.65	.09	4.491	6
14	τ^5	15 31.4	73 31	5.72	.05	5.366	6
15	τ^6	15 35.9	73 38	6.08	.05	4.500	6
16	χ	15 36.6	76 47	5.00	0.08	4.502	6
17	ι	15 36.7	69 59	4.59	.09	2.467	5...4
18	τ^7	15 37.0	71 11	6.04	.02	4.502	6
19	ψ	15 38.5	87 8	5.83	.06	4.500	6
20	α	15 38.9	83 14	2.67	.02	2.467	2...3
21	τ^8	15 39.7	72 23	5.79	0.05	4.500	6
22	25 (Λ^2)	15 40.4	91 28	5.00	.08	4.489	6
23	β	15 40.1	74 14	3.55	.13	2.467	3...4
24	λ	15 41.1	82 18	4.68	.04	...	4...5
25	υ	15 42.2	75 33	5.77	.05	4.494	6

Nos. 2, 6, and 22. Spectrum I a ! (Vogel.)

No. 3. Σ 1930, Dist. 10″. Observed as one mass.

Nos. 4 and 8. Called red in Uranometria Argentina.

No. 4. Double. Dist. 2″.3. Observed as one mass.

Nos. 5 and 12. Spectrum II a ! (Vogel.)

No. 9. Σ 1954, Dist. 3″.5. Binary. Observed as one mass. Spectrum I a ! (Vogel.)

No. 13. Birm., No. 356. The colour is orange. Variable? (See Birm., p. 279.) Spectrum III a !!! (Vogel.)

No. 20. Spectrum II a !!! (Vogel.)

No. 23. Σ 1970, Dist. 31″. The brighter star observed. This magnitude, 3.55, was also given by a second determination made on 1885.366. Spectrum I a ! (Vogel.)

No. 24. 4.68 is the mean of two determinations (4.75, 4.61) made on 1882.467 with twenty extinctions, and on 1885.366 with six. H.P.=4.35.

Reference Number.	Star's Designation.	R.A. 1890.	N.P.D. 1890.	Adopted Zenithal Magnitude. Polaris 2.05.	Average Deviation in Magnitude.	Date, 1880 +	Mag. Argel. Uran.
		h. m.	° '				
26	κ	15 43.8	71 31	3.94	0.10	...	4
27	μ	15 43.9	93 6	3.30	.17	2.467	3...4
28	ω	15 44.7	87 28	5.05	.07	4.500	6
29	ε.................	15 45.3	85 11	3.65	.05	2.467	3...4
30	36 (h)	15 45.5	92 45	5.31	.05	4.489	5
31	ρ.................	15 46.4	68 41	5.00	0.04	4.500	5
32	P. xv. 212	15 49.7	69 22	6.21	.03	4.500	6
33	P. xv. 215	15 50.7	71 3	6.40	.07	4.502	6
34	γ	15 51.4	73 58	3.83	.07	2.467	4...3
35	φ	15 52.2	75 16	5.46	.11	4.500	6
36	π	15 57.6	66 53	4.48	0.08	...	5...4
37	45	16 2.4	79 49	5.65	.07	4.502	6
38	47	16 3.2	81 10	6.00	.09	4.502	6
39	σ	16 16.5	88 43	4.92	.06	4.500	5
40	ν....................	17 14.6	102 44	...	.03	2.472	5...4
41	ξ	17 31.3	105 20		0.14	2.472	4...3
42	ο..................	17 35.2	102 49	...	.07	2.472	5...4
43	ζ..................	17 54.7	93 41	4.48	.06	4.502	5
44	η	18 15.6	92 56	3.50	.10	...	3
45	59	18 21.6	89 52	5.50	.08	...	6

No. 26. Birm., No. 358. The colour is orange. 3.94 is the mean of two determinations (3.96, 3.92) made on 1882.467 with twenty extinctions, and on 1885.366 with six. Spectrum III a !! (Vogel.)

Nos. 28 and 38. Called red in the Uranometria Argentina.

Nos. 29, 37, and 39. Spectrum I a ! (Vogel.)

No. 31. Birm., No. 363. The red colour is not salient.

No. 36. 4.48 is the mean of two determinations (4.40, 4.56) made on 1882.472 with twenty extinctions, and on 1885.366 with six. H.P. = 4.99.

No. 40. Σ Unnum., Dist. 51″. The resulting magnitude after applying a mean absorption correction is 4.32.

Nos. 41 and 42. The resulting magnitude of these stars after applying a mean absorption correction is respectively 3.34 and 4.28.

No. 44. 3.50 is the mean of two determinations (3.65, 3.48) made on 1882.472 with twenty extinctions, and on 1885.366 with six. H.P. = 3.35.

No. 45. Σ 2316, Dist. 4″. Observed as one mass. A second determination of this star made on 1885.366 gave the same magnitude 5.50. H.P. = 5.21.

Reference Number.	Star's Designation.	R.A. 1890.	N.P.D. 1890.	Adopted Zenithal Magnitude. Polaris 2.05.	Average Deviation in Magnitude.	Date, 1880 +	Mag. Argel. Uran.
		h. m.	° ′				
46	60	18 24.0	92 3	5.61	0.06	4.500	6
47	e	18 31.9	90 25	5.84	.04	4.502	6
48	62	18 50.1	83 31	5.91	.09	4.500	6
49	θ1	18 50.7	86 2	3.91	.06	2.472	} 4...3
50	θ2	18 50.7	86 2	4.23	.03	2.472	
51	64	18 51.7	87 37	5.66	0.06	4.500	6

Nos. 46 and 48. Called red in Uranometria Argentina.
No. 47. Spectrum I a ! (Vogel.)
Nos. 49 and 50. These stars form Σ 2417, Dist. 22″. Strong suspicion of variability. (See Uran. Argent., p. 322.) Spectrum I a ! (Vogel.)

SEXTANS.

1	1	9 31.4	82 40	5.08	0.06	4.264	6
2	2	9 32.6	84 51	4.61	.04	4.264	5
3	B.A.C. 3336	9 40.4	82 47	5.88	.02	4.264	6
4	P. xix. 171	9 40.7	87 42	5.70	.05	4.272	6
5	7	9 46.5	87 2	6.14	.04	4.272	6
6	8	9 47.1	97 35	5.34	0.07	4.248	5
7	12	9 54.0	86 5	6.73	.02	4.272	6
8	13	9 58.4	86 15	6.70	.02	4.272	6
9	15	10 2.3	89 50	4.89	.04	...	4...5
10	17	10 4.7	97 52	6.10	.04	4.231	
11	18	10 5.5	97 52	5.70	0.02	4.231	6
12	19	10 7.1	84 51	6.04	.06	4.248	6
13	22	10 12.2	97 31	5.47	.03	4.248	6
14	23	10 15.4	87 9	6.79	.07	4.272	6
15	29	10 23.9	92 11	5.15	.04	4.215	5
16	30	10 24.7	90 4	5.02	0.04	4.220	5
17	35	10 37.6	84 40	6.15	.05	4.264	6
18	41	10 44.8	98 19	5.70	.03	4.272	5

No. 1. = 10 Leonis. (See Introduction to B.A.C., p. 75.)
No. 3. B.A.C. assigns this star to Leo. Nos. 5 and 15. Spectrum I a ! (Vogel.)
No. 9. 4.89 is the mean of six accordant determinations made on as many nights, one at Oxford and five at Cairo, involving seventy extinctions, made between 1882.360 and 1883.108. H.P.=4.51.
No. 11. Birm., No. 234. The red colour is not salient.
No. 17. Σ 1466, Dist. 7″. Observed as one mass.

TAURUS.

Reference Number.	Star's Designation.	R.A. 1890.	N.P.D. 1890.	Adopted Zenithal Magnitude. Polaris 2.05.	Average Deviation in Magnitude.	Date, 1880 +	Mag. Argel. Uran.
		h. m.	° ′				
1	P. iii. 6	3 6.6	83 45	5.95	0.10	4.861	6
2	0	3 18.9	81 21	3.56	.06	2.877	4...3
3	ξ	3 21.2	80 39	3.72	.03	2.877	4...3
4	4	3 24.7	79 2	5.31	.06	4 872	5
5	Σ 401	3 24.7	62 49	5.93	.06	4.845	6
6	5	3 24.8	77 26	4.15	0.10	2.869	4
7	6	3 26.7	81 0	5.70	.05	4.861	6
8	7	3 27.9	65 54	6.01	.03	4.845	6
9	10	3 31.3	89 57	4.40	.09	2.877	4...5
10	P. iii. 103	3 33.2	73 49	6.19	.08	4.853	6
11	12	3 34.2	87 18	5.99	0.10	4.861	6
12	13	3 36.0	70 39	5.49	.04	4.872	6...5
13	P. iii. 128	3 38.1	69 25	5.93	.04	4.845	6
14	29	3 39.8	84 18	5.09	.05	4.861	6...5
15	30	3 42.2	79 12	5.09	.02	4.861	5
16	P. iii. 170	3 43.7	64 45	5.56	0.07	4.853	6
17	D.M.+21°, No. 539	3 45.2	68 17	5.99	.09	...	6
18	D.M.+12°, No. 516	3 45.2	77 17	6.10	.06	4.890	6
19	31	3 46.1	83 48	5.87	.05	4.883	6
20	P. iii. 187	3 46.9	73 0	6.17	.03	4.872	6
21	32	3 50.4	67 50	5.77	0.09	4.845	6
22	P. iii. 203	3 51.2	84 17	6.09	.04	4.890	6
23	P. iii. 215	3 54.5	72 7	6.07	.04	...	6
24	λ	3 54.6	77 49	3.43	.05	2.877	3...4
25	Bradley 547	3 54.7	70 6	6.75	.07	4.853	6

No. 1. B.A.C. assigns this star to Cetus.
No. 2. Called red in Uranometria Argentina. Auwers thought the star variable with a period of seven days. (See Ast. Nach., vol. l, p. 105.)
Nos. 3 and 4. Spectrum I a !! (Vogel.)
No. 5. Σ 401, Dist. 11″. Observed as one mass.
No. 6. Spectrum II a ! (Vogel.)   Nos. 7, 14, 15, and 22. Spectrum I a ! (Vogel.)
No. 8. Σ 412, Dist.<1″. Observed as one mass.
No. 15. Σ 452, Dist. 9″. Observed as one mass.
No. 17. 5.99 is the mean of two determinations (5.97, 6.00) made on 1884.872 and on 1885.168, each with six extinctions. H.P.=5.68.
No. 23. 6.07 is the mean of two determinations (5.98, 6.15) made on 1884.890 and on 1885.168, each with six extinctions. H.P.=5.68.
No. 24. Variability discovered by Baxendell in 1848. Max. 3.4, Min. 4 2 mag. Period 3ᵈ, 22ʰ 52ᵐ. Epoch of minimum 1866, Dec. 31, 12ʰ 34ᵐ. Spectrum I a ! (Vogel.)

Reference Number.	Star's Designation.	R.A. 1890.	N.P.D. 1890.	Adopted Zenithal Magnitude Polaris − 2.05.	Average Deviation in Magnitude.	Date, 1880+	Mag. Argel. Uran.
		h. m.	° ′				
26	P. lii. 220	3 55.8	80 19	5.98	0.05	4.861	6
27	ν...................	3 57.3	84 19	4.10	.04	2.874	4
28	36	3 57.8	66 12	5.71	.04	4.845	6
29	P. iii. 234	3 58.0	82 6	5.80	.05	...	6
30	40	3 58.0	84 52	5.56	.03	4.880	6
31	37	3 58.2	68 13	4.84	0.09	...	5...4
32	P. iii. 238	3 58.5	87 28	5.71	.06	4.960	6
33	41	3 59.9	62 42	5.36	.09	4.845	6...5
34	ψ	4 0.2	61 18	5.27	.05	4.853	6
35	43	4 2.8	70 41	6.07	.02	4.872	6
36	44 (ν)	4 4.1	63 48	5.78	0.05	4.853	6
37	45	4 5.5	84 46	5.86	.06	4.883	6
38	46	4 7.6	82 34	5.43	.04	4.872	6
39	47	4 8.0	81 1	5.10	.04	4.960	5
40	P. iv. 19	4 8.6	80 17	5.27	.04	4.880	6...5
41	48	4 9.5	74 52	6.19	0.05	4.960	6
42	μ	4 9.6	81 23	4.30	.05	5.248	4...5
43	P. iv. 24...........	4 9.6	84 5	6.95	.02	4.962	} 6
44	P. iv. 25...........	4 9.7	84 6	6.25	.04	4.962	
45	ω	4 10.8	69 41	5.08	.03	...	6...5
46	53	4 13.0	69 7	5.76	0.07	4.853	6
47	56	4 13.1	68 29	5.65	.02	4.890	6
48	γ	4 13.5	74 38	3.55	.10	...	4
49	φ	4 13.6	62 55	5.21	.03	4.853	5...6
50	57 (h)	4 13.8	76 14	5.77	.08	4.960	6

No. 26. Called red in Uranometria Argentina. Spectrum I a ! (Vogel.)

No. 27. 4.10 is the mean of two determinations (4.06, 4.14) made on 1882.869 and on 1882.877, each with twenty extinctions. Spectrum I a !! (Vogel.)

No. 29. 5.80 is the mean of two determinations (5.87, 5.73) made on 1884.890 and on 1885.168, each with six extinctions. H.P.=5.55.

No. 31. 4.84 is the mean of two determinations (4.88, 4.80) made on 1882.869 with twenty extinctions, and on 1885.248 with six. H.P.=4.45.

Nos. 38 and 39. Spectrum I a ! (Vogel.)

Nos. 39 and 40. Called red in Uranometria Argentina.

No. 41. Variable? Schmidt. (See Ast. Nach., vol. lxxx, p. 253 and p. 383.)

No. 42. The determination of 1882.897 (see Memoirs R.A.S., vol. xlvii, p. 451) is rejected. The night was hazy.

No. 45. 5.08 is the mean of two determinations (5.05, 5.10) made on 1884.853 and on 1885.168, each with six extinctions. H.P.=4.65.

No. 48. 3.55 is the mean of two determinations (3.50, 3.60) made on 1882.877 with twenty extinctions, and on 1885.168 with six. Spectrum II a ! (Vogel.)

No. 50. Spectrum I a ! (Vogel.)

Reference Number.	Star's Designation.	R.A. 1890.	N.P.D. 1890.	Adopted Zenithal Magnitude. Polaris 2.05.	Average Deviation in Magnitude.	Date, 1880 +	Mag. Argel. Uran.
		h. m.	° ′				
51	P. iv. 49	4 14.9	84 8	6.02	0.06	4.890	6
52	χ	4 15.9	64 38	5.69	.03	4.853	6...5
53	60	4 15.9	76 11	5.92	.10	4.883	6
54	δ¹	4 16.6	72 43	3.90	.08	5.168	4
55	P. iv. 61	4 17.0	69 17	6.05	.03	4.890	6
56	62	4 17.4	65 57	6.11	0.03	4.960	6
57	64	4 17.7	72 49	4.97	.06	4.890	6
58	66 (r)..............	4 17.9	80 48	5.21	.04	4.880	5...6
59	κ	4 18.8	67 57	4.60	.07	2.869	5...4
60	68	4 19.1	72 19	4.48	.02	4.960	5
61	υ	4 19.7	67 26	4.64	0.07	2.869	5...4
62	71	4 20.1	74 38	5.03	.04	...	6
63	π	4 20.4	75 32	5.09	.11	4.880	5
64	72	4 20.7	67 15	5.65	.05	4.968	6
65	P. iv. 82	4 21.5	68 37	6.00	.04	...	6
66	D.M.+10°. No. 577	4 21.5	79 2	6.07	0.03	4.883	6
67	ε.....................	4 22.2	71 4	3.69	.04	5.248	4...3
68	θ¹	4 22.3	74 17	4.13	.04	2.902	4...5
69	θ²	4 22.4	74 22	3.65	.04	2.902	4...5
70	79 (b)	4 22.7	77 12	5.06	.07	4.880	6...5
71	P. iv. 99............	4 24.3	74 3	5.09	0.05	4.968	5
72	83	4 24.4	76 31	5.62	.06	4.984	6
73	ρ	4 27.6	75 23	5.01	.07	4.880	5
74	P. iv. 111	4 27.8	61 16	5.70	.03	4.984	6
75	D.M.+5°. No. 679	4 28.3	84 40	5.97	.03	4.968	6

No. 52. Σ 528, Dist. 19″. The bright star observed.

Nos. 53, 57, and 58. Spectrum I a! (Vogel.)

No. 54. The determination made on 1882.877 is rejected. (See Memoirs R.A.S., vol. xlvii, p. 451.) The night was hazy and variable.

No. 56. Σ 534, Dist. 29″. The bright star observed.

No. 59. Σ 9¹. The very distant companion (67 Tauri) is not observed.

No. 60. Spectrum I a !! (Vogel.)

No. 62. 5.03 is the mean of two determinations (4.99, 5.07) made on 1884.968 and on 1885.174, each with six extinctions. H.P.=4.61.

No. 65. 6.00 is the mean of two determinations (5.96, 6.03) made on 1884.890 and on 1885.174, each with six extinctions. H.P.=5.68.

No. 67. The determination made on 1882.877 is rejected. (See Memoirs R.A.S., vol. xlvii, p. 451.) See note to No. 54.

Nos. 68 and 69. These stars form Σ 10¹, Dist. 337″.

Nos. 69 and 70. Spectrum I a !! (Vogel.)

Nos. 71, 72, 73, and 75. Spectrum I a ! (Vogel.)

Reference Number.	Star's Designation.	R.A. 1890.	N.P.D. 1890.	Adopted Zenithal Magnitude. Polaris 2.05.	Average Deviation in Magnitude.	Date, 1880+	Mag. Argel. Uran.
		h. m.	° ′				
76	α	4 29.6	73 43	1.12	0.05	...	1
77	88	4 29.6	80 4	4.47	.05	2.902	5...4
78	W.B. iv. 650......	4 31.8	69 32	5.88	.10	4.984	6
79	90	4 32.0	77 43	4.57	.06	2.902	5...4
80	σ¹	4 33.0	74 25	5.34	.09	4.890	} 5
81	σ²	4 33.0	74 18	5.06	0.06	4.890	
82	P. iv. 146	4 33.2	82 21	5.72	.04	4.968	6
83	93	4 33.9	78 1	5.50	.06	4.960	6...5
84	P. iv. 148	4 34.4	61 36	6.00	.06	4.968	6
85	τ	4 35.6	67 15	4.65	.06	2.902	4...5
86	97	4 44.9	71 21	5.20	0.10	4.984	5...6
87	Bradley 686	4 51.0	73 1	5.74	.01	4.984	6
88	Bradley 684	4 51.1	66 13	6.00	.02	4.968	6
89	98	4 51.4	65 7	5.79	.05	4.960	6...5
90	ι...................	4 56.5	68 34	4.90	.04	4.890	5
91	104................	5 0.9	71 30	5.30	0.07	4.880	5...6
92	106................	5 1.3	69 44	5.47	.05	4.984	6...5
93	105................	5 1.4	68 26	5.67	.06	5.000	6
94	103................	5 1.4	65 54	5.69	.10	4.984	6
95	109................	5 12.7	68 1	5.44	.05	...	6
96	111................	5 18.0	72 43	5.26	0.05	4.968	6...5
97	β	5 19.3	61 29	1.79	.07	...	2
98	115................	5 20.8	72 8	5.69	.05	...	6
99	114................	5 21.0	68 9	5.06	.05	4.984	6
100	116................	5 21.4	74 13	5.62	.03	4.960	6

No. 76. Birm., No. 81. The colour is yellow. Σ 2¹¹. The faint distant companion not observed. Seidel suggests variability. (See Result. Phot. Mess., p. 162.) Selected by Dr. G. Müller to detect the effect of atmospheric absorption on coloured stars. See Photometrische Untersuchungen. Spectrum II a !!! (Vogel.) 1.12 is the mean of seven determinations made on as many nights, four at Oxford and three at Cairo, involving eighty extinctions, between 1882.902 and 1883.168.

Nos. 77, 79, and 82. Spectrum I a ! (Vogel.)

Nos. 80 and 81. These stars form Σ 11¹, Dist. 428″. Spectrum I a ! (Vogel.)

No. 85. Double. Dist. 61″. The distant companion not observed.

No. 86. Spectrum I a ! (Vogel.)

No. 95. 5.44 is the mean of two determinations (5.40, 5.47) made on 1884.962 and on 1885.174, each with six extinctions. H.P. = 5.15.

No. 97. β Tauri = γ Aurigæ. (See Introduction to B.A.C., p. 75.) 1.79 is the mean of five determinations, made on as many nights, three at Oxford and two at Cairo, involving sixty extinctions, between 1882.902 and 1883.193.

No. 98. 5.69 is the mean of two determinations (5.64, 5.74) made on 1884.984 and on 1885.174, each with six extinctions. H.P. = 5.38.

Refer-ence Number.	Star's Designation.	R.A. 1890.	N.P.D. 1890.	Adopted Zenithal Magnitude. Polaris 2.05.	Average Deviation in Magnitude.	Date, 1880 +	Mag. Argel. Uran.
		h. m.	° ′				
101	118..............	5 22.5	64 56	5.48	0.06	4.984	6
102	119..............	5 25.8	71 29	4.60	.10	4.962	6...5
103	Σ 730..............	5 25.9	73 2	5.67	.09	5.000	6
104	121..............	5 28.7	66 2	5.77	.04	...	6
105	122..............	5 30.7	73 2	5.60	.04	4.960	6
106	ζ..............	5 31.1	68 55	3.00	0.06	2.902	3...4
107	125..............	5 32.9	64 10	5.07	.07	5.000	6
108	126..............	5 34.9	73 31	5.10	.05	5.000	5
109	P. v. 192	5 36.7	66 51	0.04	.05	4.968	6
110	130..............	5 41.0	72 19	5.48	.02	4.960	6
111	133..............	5 41.5	76 8	5.27	0.02	4.883	6
112	132..............	5 42.3	65 28	5.30	.03	4.968	5...6
113	134..............	5 43.4	77 23	4.98	.02	4.962	5...6
114	135..............	5 44.2	75 44	5.73	.02	4.984	6
115	137..............	5 46.1	75 51	5.67	.08	4.960	6
116	136..............	5 46.4	62 25	4.66	0.09	...	5
117	139..............	5 51.2	64 4	5.00	.04	4.968	5...6

No. 101. Σ 716, Dist. 5″. Observed as one mass.
No. 102. Birm., No. 111. The colour is distinctly red. Spectrum III a ! ! (Vogel.)
No. 103. Σ 730, Dist. 10″. Observed as one mass.
No. 104. **5.77** is the mean of two determinations (5.75, 5.80) made on 1884.968 and on 1885.174, each with six extinctions. **H.P. = 5.43.**
Nos. 113 and 115. Spectrum I a ! (Vogel.)
No. 116. **4.66** is the mean of two determinations (4.72, 4.60) made on 1885.000 and on 1885.174, each with six extinctions. **H.P. = 4.46.**

TRIANGULUM.

Reference Number.	Star's Designation.	R.A. 1890.	N.P.D. 1890.	Adopted Zenithal Magnitude. Polaris 2.05.	Average Deviation in Magnitude.	Date, 1880+	Mag. Argel. Uran.
		h. m.	o ′				
1	B.A.C. 514	1 35.4	60 30	6.07	0.03	4.984	6
2	B.A.C. 516	1 35.7	55 19	5.66	.07	...	6
3	P. i. 171	1 42.4	57 53	5.95	.06	4.999	6
4	a	1 46.8	60 59	3.50	.06	2.656	4...3
5	ε.................	1 56.5	57 15	5.63	.04	5.022	5...6
6	β	2 3.0	55 32	3.12	0.06	2.656	3
7	6.................	2 6.0	60 13	5.50	.05	4.984	6...5
8	7.................	2 9.4	57 9	5.28	.07	4.999	5
9	δ.................	2 10.3	56 17	5.17	.01	5.022	6...5
10	γ	2 10.8	56 40	4.35	.06	2.656	4...5
11	10	2 12.6	61 52	5.55	0.06	...	6
12	11	2 20.9	58 41	5.60	.08	4.999	6
13	12	2 21.7	60 49	5.54	.02	4.984	6...5
14	14	2 25.4	54 19	5.43	.07	4.999	6
15	15	2 29.1	55 48	5.65	.01	4.984	6...5

No. 2. B.A.C. assigns this star to Andromeda. 5.66 is the mean of two determinations (5.65, 5.68) made on 1884.999 and on 1885.137, each with six extinctions. H.P.=5.37.
No. 5. Σ 201, Dist. 5″. Observed as one mass.
No. 7. Σ 227, Dist. 3″.5. Observed as one mass.
No. 11. 5.55 is the mean of two determinations (5.57, 5.53) made on 1884.984 and on 1885.137, each with six extinctions. H.P.=5.29.
No. 15. Birm., Add. I, No. 9. The red colour is not salient. Secchi suggests variability. (See Prodromo.)

URSA MAJOR.

Reference Number.	Star's Designation.	R.A. 1890.	N.P.D. 1890.	Adopted Zenithal Magnitude. Polaris 2.05.	Average Deviation in Magnitude.	Date, 1880 +	Mag. Argel. Urau.
		h. m.	° ′				
1	B.A.C. 2707	8 1.8	21 12	5.45	0.05	5.292	6
2	B.A.C. 2765	8 9.7	27 9	5.82	.11	5.303	6
3	P. viii. 46	8 19.4	22 20	5.95	.00	5.283	6
4	o...................	8 21.2	28 55	3.36	.09	2.913	3...4
5	2...................	8 24.7	24 29	5.48	.02	5.305	5
6	B.A.C. 2887	8 30.2	36 13	5.78	0.01	5.283	6
7	π^2	8 30.6	25 17	4.76	.03	2.913	5...4
8	P. viii. 105.........	8 31.1	36 54	6.10	.10	5.292	6
9	P. viii. 137.........	8 38.9	22 53	6.05	.14	5.283	6
10	5...................	8 44.3	27 38	5.57	.10	5.303	5
11	6...................	8 47.2	24 58	5.71	0.07	5.292	6
12	ι	8 51.7	41 31	3.23	.07	...	3
13	ρ...................	8 52.6	21 57	4.99	.13	5.283	5
14	10	8 53.5	47 46	4.10	.09	2.913	4
15	B.A.C. 3072	8 55.9	35 17	5.71	.07	5.305	6
16	κ...................	8 56.1	42 24	3.62	0.03	2.913	3...4
17	B.A.C. 3086	8 58.2	30 12	6.23	.07	5.283	6
18	σ^1	8 58.7	22 41	5.26	.12	5.303	5
19	P. viii. 245.........	8 59.6	51 6	4.74	.04	2.938	5
20	σ^2	9 0.7	22 25	5.00	.13	5.303	5
21	15	9 1.1	37 57	4.68	0.10	5.308	5
22	τ...................	9 1.8	26 2	4.94	.15	5.283	5...4
23	B.A.C. 3116	9 4.8	16 36	5.83	.16	5.292	6
24	16	9 6.0	28 8	5.14	.07	5.308	5
25	17	9 7.7	32 48	5.51	.11	5.292	6

No. 1. Called 55 Camelopardali in the B.A.C.

No. 2. B.A.C. assigns this star to Camelopardalus.

No. 5. Σ Unnum. The distant star not observed.

Nos. 6, 8, 14, and 19. B.A.C. assigns these stars to Lynx.

No. 12. OΣ 196, Dist. 10″. Observed as one mass. 3.23 is the mean of four determinations made on as many nights, two at Oxford and two at Cairo, involving fifty extinctions, made between 1882.913 and 1883.089.

No. 18. Birm., Add. I, No. 39. The colour is slightly orange.

No. 20. Σ 1306, Dist. 3″. Observed as one mass.

Reference Number.	Star's Designation.	R.A. 1890.	N.P.D. 1890.	Adopted Zenithal Magnitude. Polaris 2.05.	Average Deviation in Magnitude.	Date, 1880 +	Mag. Argel. Uran.
		h. m.	° ′				
26	18	9 8.3	35 32	5.00	0.07	5.292	5
27	P. ix. 19	9 10.1	42 43	5.79	.06	5.283	6
28	B.A.C. 3172	9 13.6	32 50	5.86	.08	5.283	6
29	P. ix. 78	9 21.5	43 55	5.35	.03	5.292	6
30	23	9 22.8	26 27	3.66	.08	2.913	3...4
31	22	9 24.5	17 18	5.88	0.11	5.283	6
32	24	9 24.8	19 41	4.92	.06	...	5...4
33	θ	9 25.6	37 49	3.12	.06	2.913	3
34	26	9 27.3	37 27	4.75	.07	5.303	5
35	27	9 32.8	17 15	5.44	.08	5.283	6
36	B.A.C. 3287	9 32.9	20 16	5.62	0.01	5.305	6
37	P. ix. 159	9 38.7	32 22	5.42	.12	5.292	6
38	P. ix. 169	9 41.5	43 28	5.39	.04	5.308	5
39	υ..................	9 43.2	30 26	3.94	.10	2.913	4...3
40	φ	9 44.6	35 25	4.73	.11	...	5...4
41	31	9 48.5	39 40	5.15	0.08	5.292	5
42	P. ix. 201	9 49.6	32 3	5.75	.07	5.283	6
43	B.A.C. 3402	9 52.3	32 40	5.53	.10	5.283	6
44	P. ix. 229	9 57.3	35 34	5.90	.04	5.292	6
45	R. 2460	10 7.5	29 28	5.98	.08	5.303	6
46	32	10 10.0	24 20	5.78	0.09	5.283	6
47	λ	10 10.5	46 32	3.52	.03	2.938	3...4
48	P. x. 26	10 12.7	20 42	5.88	.05	5.308	6
49	L. 19985	10 13.1	35 38	6.56	.10	5.303	6
50	μ	10 15.8	47 57	3.12	.05	2.913	3
51	P. x. 42	10 16.2	23 52	5.01	0.03	5.305	5
52	B.A.C. 3567	10 21.2	40 37	6.17	.04	5.303	6
53	R. 2498	10 22.8	25 11	6.03	.05	5.308	6
54	36	10 23.6	33 27	4.98	.09	5.308	5
55	B.A.C. 3607	10 26.8	49 0	5.11	.12	5.193	5

No. 29. B.A.C. assigns this star to Lynx.

No. 30. Σ 1351, Dist. 23″. The brighter star observed.

No. 32. 4.92 is the mean of two determinations (4.99, 4.86) made on 1882.913 with twenty extinctions, and on 1885.415 with six. H.P.=4.59.

No. 38. B.A.C. assigns this star to Leo Minor.

No. 40. OΣ 208. 4.73 is the mean of two determinations (4.80, 4.67) made on 1882.938 with twenty extinctions, and on 1885.415 with six. H.P.=4.43.

No. 50. Birm., No. 238. The colour is orange.

Reference Number.	Star's Designation.	R.A. 1890.	N.P.D. 1890.	Adopted Zenithal Magnitude. Polaris 2.05.	Average Deviation in Magnitude.	Date, 1880 +	Mag. Argel. Uran.
		h. m.	° ′				
56	37	10 28.1	32 21	.5.38	0.04	5.303	5
57	B.A.C. 3639	10 32.8	35 45	5.71	.03	5.308	6
58	B.A.C. 3645	10 34.0	20 59	5.78	.07	5.311	6
59	38	10 34.5	23 42	5.00	.03	5.311	5
60	P. x. 126	10 35.2	20 21	4.96	.09	...	5
61	39	10 36.8	32 13	5.80	0.02	5.303	6
62	P. x. 135	10 37.1	43 13	5.13	.05	5.308	5
63	D.M. 65°, No. 803	10 41.4	24 17	5.99	.07	...	6
64	43	10 44.4	32 50	6.02	.08	5.305	6
65	42	10 44.5	30 6	5.87	.12	5.305	6
66	R. 2571	10 45.9	36 51	6.66	0.10	5.311	} 6
67	R. 2572	10 46.0	36 55	6.44	.06	5.311	
68	R. 2569	10 46.1	19 34	5.90	.05	5.308	6
69	44	10 46.9	34 50	5.56	.05	5.305	5
70	ω	10 47.7	46 13	4.91	.09	5.193	5
71	O.A. 11292	10 52.7	37 31	6.34	0.10	5.308	6
72	47	10 53.3	48 59	5.23	.01	5.196	5
73	P. x. 203	10 53.3	53 17	6.14	.06	5.206	6
74	B.A.C. 3758	10 53.9	43 53	5.88	.06	5.193	6
75	B.A.C. 3760	10 54.1	46 29	5.90	.09	5.193	6
76	49	10 54.7	50 12	5.18	0.05	5.196	5
77	β	10 55.2	33 1	2.17	.06	...	2...3
78	a	10 57.0	27 39	1.89	.04	...	2
79	51	10 58.4	51 10	5.87	.10	5.193	6
80	O.A. 11453	11 2.6	22 11	5.91	.10	5.308	6

No. 60.   4.96 is the mean of two determinations (4.94, 4.97) made on 1885.305 with twenty extinctions, and on 1885.423 with six. H.P.=5.25.

No. 63.   5.99 is the mean of two determinations (6.00, 5.98) made on 1885.308 with twenty extinctions, and on 1885.423 with six. H.P. = 6.32.

No. 77.   2.17 is the mean of the determinations made on ten different nights involving one hundred and ten extinctions, details of which are given in the Preface.

No. 78.   Birm., No. 252. The colour is supposed to be variable. (See Ast. Nach., vol. lxxxviii, p. 363.) 1.89 is the mean of the determinations made on ten different nights involving one hundred and ten extinctions (see Preface). The variability of this star and of the other brighter stars in Ursa Major has been discussed by Schmidt. (See Ast. Nach., vol. xlvi, p. 299.) Argelander distrusts the evidence in favour of variability. (See Bonn. Beobacht., vol. vii, pp. 401 and 515.)

Reference Number.	Star's Designation.	R.A. 1890.	N.P.D. 1890.	Adopted Zenithal Magnitude. Polaris 2.05.	Average Deviation in Magnitude.	Date, 1880 +	Mag. Argel. Uran.
		h. m.	° ′				
81	P. x. 252	11 3.3	53 6	5.96	0.07	5.196	6
82	ψ	11 3.5	44 54	3.21	.05	2.913	3
83	B.A.C. 3821	11 5.1	21 7	6.06	.03	...	6
84	R. 2648	11 9.7	36 37	6.16	.03	5.308	6
85	P. xi. 19............	11 10.5	39 55	5.76	.13	5.308	6
86	ξ..................	11 12.3	57 51	3.75	0.07	2.957	4...3
87	ν..................	11 12.5	56 18	3.49	.09	...	3...4
88	55	11 13.1	51 14	4.91	.03	5.196	5
89	R. 2662	11 14.2	22 17	6.08	.11	...	6
90	P. xi. 43............	11 16.3	25 4	5.93	.07	5.423	6
91	56	11 17.8	45 55	5.27	0.16	5.196	6
92	P. xi. 59............	11 19.9	33 33	5.71	.15	5.308	6
93	57	11 23.2	50 3	5.10	.02	5.206	5
94	R. 2693	11 23.6	32 39	5.88	.07	5.308	6
95	58	11 24.6	46 13	5.95	.03	5.196	6
96	B.A.C. 3918	11 26.1	28 18	5.53	0.01	5.308	6
97	B.A.C. 3931	11 29.0	34 36	5.67	.07	5.210	6
98	B.A.C. 3949	11 31.9	38 46	5.85	.14	5.311	6
99	59	11 32.5	45 46	5.53	.05	5.196	6
100	R. 2721	11 32.6	25 2	6.28	.05	5.308	6

No. 82. Birm., No. 254. The colour is yellow. This star has been suspected of variability. (See Ast. Nach., vol. xlvi, p. 303.)

No. 83. B.A.C. assigns this star to Draco. 6.06 is the mean of two determinations (6.06, 6.05) made on 1885.305 with twenty extinctions, and on 1885.423 with six. H.P. = 6.57.

No. 84. Σ 1520, Dist. 13″. Observed as one mass.

No. 86. Σ 1523. Binary. Period 60.80 years. Semi-axis major 2″.58. (See Oxford University Astron. Obs., No. 1, p. 62.) Observed as one mass.

No. 87. Σ 1524, Dist. 8″. Observed as one mass. Birm., No. 259. The colour is orange. 3.49 is the mean of two determinations (3.50, 3.47) made on 1882.957 with twenty extinctions, and on 1885.415 with six.

No. 89. 6.08 is the mean of two determinations (6.08, 6.07) made on 1885.305 and on 1885.423, each with six extinctions.

No. 93. Σ 1543, Dist. 6″. Observed as one mass.

No. 96. OΣ 235, Dist. 1″. Rapid binary. Observed as one mass.

No. 100. Σ 1559, Dist. 2″. Observed as one mass.

Refer-ence Number.	Star's Designation.	R.A. 1890.	N.P.D. 1890.	Adopted Zenithal Magnitude. Polaris 2.05.	Average Deviation in Magnitude.	Date, 1880+	Mag. Argel. Uran.
		h. m.	o ′				
101	61	11 35.3	55 10	5.55	0.10	...	5
102	62	11 35.9	57 38	5.60	.17	5.196	6
103	χ	11 40.3	41 36	3.80	.07	2.957	4
104	B.A.C. 3985	11 41.0	33 45	5.58	.06	...	6
105	P. xi. 164	11 44.0	54 27	5.58	.06	5.210	6
106	D.M. 34°, No. 2264	11 45.4	56 0	6.22	0.15	5.196	6
107	γ	11 48.0	35 42	2.30	.09	...	2...3
108	} 65 ....{	11 49.4	42 54	6.32	.07	5.311	} 6
109		11 49.5	42 55	6.86	.07	5.311	
110	66	11 50.2	32 47	5.86	.04	5.308	6
111	B.A.C. 4036	11 51.2	27 50	6.06	0.01	5.308	6
112	R. 2771	11 51.6	49 2	6.23	.18	5.196	6
113	W.B. xi. 1066 ...	11 56.0	53 19	5.85	.03	5.203	6
114	67	11 56.6	46 20	5.09	.12	5.196	5
115	B.A.C. 4074	12 0.1	26 27	6.21	.01	5.311	6
116	R. 2800	12 3.8	35 59	6.36	0.11	5.196	6
117	δ....................	12 10.0	32 21	3.41	.06	...	3...4
118	70	12 15.5	31 31	5.58	.07	5.210	6
119	73	12 22.4	33 41	5.75	.10	5.196	6
120	74	12 24.8	30 59	5.62	.05	5.203	6
121	76	12 36.8	26 41	5.99	0.05	5.196	6
122	R. 2913	12 42.6	26 37	5.90	.14	5.210	6
123	ε....................	12 49.2	33 27	1.80	.07	...	2
124	B.A.C. 4348	12 51.5	35 20	5.82	.01	5.196	6
125	78	12 56.0	33 2	4.98	.09	5.210	6

No. 101. 5.55 is the mean of two determinations (5.53, 5.58) made on 1885.203 and on
1885.423, each with six extinctions.
No. 103. Birm., No. 266. The red colour is not salient.
No. 104. 5.58 is the mean of two determinations (5.56, 5.60) made on 1885.308 and on
1885.423, each with six extinctions. H.P. = 5.33.
No. 107. 2.30 is the mean of the determinations made on twelve different nights involving
one hundred and thirty extinctions. (See Preface.)
No. 108. Σ 1579, Dist. 4″. Observed as one mass.
Nos. 108 and 109. These stars form Σ 20¹, Dist. 63″.5.
No. 116. B.A.C. assigns this star to Canes Venatici.
No. 117. 3.41 is the mean of the determinations made on twelve different nights involving
one hundred and thirty extinctions. (See Preface.) Schmidt believes this star to be slightly
variable. (See Ast. Nach., vol. lxxii, p. 343, and vol. lxxiii, p. 272.)
No. 123. 1.80 is the mean of the determinations made on ten different nights involving one
hundred and ten extinctions. (See Preface.)
No. 124. Σ 1695, Dist. 3.″ Observed as one mass.

Reference Number.	Star's Designation.	R.A. 1890.	N.P.D. 1890.	Adopted Zenithal Magnitude. Polaris 2.05.	Average Deviation in Magnitude.	Date, 1880+	Mag. Argel. Uran.
		h. m.	o ′				
126	R. 2965	13 2.0	27 22	6.29	0.03	5.203	6
127	ζ	13 19.5	34 30	2.09	.07	...	2
128	80	13 20.8	34 26	4.02	.14	5.196	5
129	R. 3025	13 24.2	38 42	6.15	.05	5.210	6
130	P. xiii. 110..........	13 24.4	29 29	5.40	.11	5.210	5...6
131	81	13 29.9	34 5	5.69	0.08	5.203	6
132	82	13 35.3	36 31	5.31	.09	5.196	6
133	R. 3072	13 36.5	32 14	6.17	.09	5.210	6
134	83	13 36.6	34 46	5.10	.11	5.203	6...5
135	R. 3081	13 39.7	37 23	5.96]	.05	5.196	6
136	R. 3089	13 41.2	33 33	6.08	0.16	5.210	6
137	84	13 42.5	35 1	5.65	.16	5.196	6
138	η	13 43.2	40 8	1.77	.05	...	2
139	R. 3104	13 46.7	30 55	6.19	.11	5.206	6
140	86	13 49.8	35 44	5.63	.10	5.196	6

No. 127. Σ 1744, Dist. 14″. Observed as one mass. 2.09 is the mean of the determinations made on ten different nights involving one hundred and ten extinctions. (See Preface.)

No. 134. Birm., No. 312. The colour is orange. Variable? Birmingham saw the star as bright as δ Ursæ Majoris in August, 1868. A determination made on 1885.423 gave the magnitude 4.95.

No. 138. 1.77 is the mean of the determinations made on ten different nights involving one hundred and ten extinctions. (See Preface.)

URSA MINOR.

Reference Number.	Star's Designation.	R.A. 1890.	N.P.D. 1890.	Adopted Zenithal Magnitude. Polaris 2.05.	Average Deviation in Magnitude.	Date, 1880 +	Mag. Argel. Uran.
		h. m.	° ′				
1	Polaris	1 18.2	1 17	(2.05)	...	...	2
2	B.A.C. 4150	12 13.7	2 57	6.12	0.02	5.210	6
3	B.A.C. 4165	12 14.3	1 41	6.15	.10	5.210	6
4	P. xiii. 109	13 23.3	17 2	5.90	.05	5.215	6
5	R. 3068	13 34.5	18 12	5.71	.03	5.215	6
6	4..................	14 9.3	11 56	4.99	0.08	5.210	5
7	B.A.C. 4732	14 10.1	20 3	5.43	.03	5.215	6...5
8	5..................	14 27.7	13 49	4.72	.05	...	5...4
9	β	14 51.0	15 24	2.26	.04	2.363	2
10	P. xiv. 260........	14 55.9	23 38	4.89	.08	5.245	5
11	B.A.C. 4989	15 2.3	23 39	5.05	0.02	5.245	6
12	B.A.C. 5058	15 13.3	22 14	5.24	.06	5.215	5...6
13	γ	15 20.9	17 46	3.02	.10	2.708	3
14	θ	15 34.7	12 17	5.02	.04	...	6...5
15	O.A. 15584	15 37.4	20 21	6.01	.03	5.210	6
16	ζ..................	15 48.0	11 52	4.05	0.07	2.708	4...5
17	R. 3524	16 7.1	12 55	5.70	.03	5.439	6
18	19	16 13.9	13 51	5.69	.03	5.245	6
19	B.A.C. 5483	16 16.3	16 20	5.94	.05	5.215	6
20	η	16 20.7	14 0	5.09	.14	5.210	5
21	P. xvi. 182........	16 31.9	10 48	5.92	0.07	...	6
22	P. xvi. 195........	16 35.4	12 20	5.97	.03	5.215	6
23	ε..................	16 57.3	7 47	4.46	.03	2.708	4...5
24	B.A.C. 5811	17 5.1	14 33	6.38	.12	5.245	6
25	R. 3727	17 27.9	9 46	5.88	.06	5.210	6

No. 1. Σ 93, Dist. 19″. The extinctions always refer to the brighter star alone. For further remarks consult the Preface.

No. 6. Birm., No. 324. The red colour is not salient.

No. 8. Birm., No. 332. The colour is orange. 4.72 is the mean of three accordant determinations made on 1882.708, 1885.245, and on 1885.439. H.P. = 4.29.

No. 9. Birm., No. 341. The red colour is not salient. Variable? (See Ast. Nach., vol. lxiv, p. 172, and vol. xlvii, p. 293, and Monthly Notices, vol. xlvii, p. 310.)

Nos. 10 and 11. The B.A.C. assigns these stars to Draco.

Nos. 10 and 14. Birm., Nos. 342 and 357 respectively. The colour is in each case slightly orange.

No. 14. 5.02 is the mean of three accordant determinations made on 1885.210, 1885.245, and on 1885.239 respectively.

No. 21. 5.92 is the mean of three accordant determinations made on 1885.210, 1885.245, and on 1885.239 respectively. H.P. = 5.54.

Reference Number.	Star's Designation.	R.A. 1890.	N.P.D. 1890.	Adopted Zenithal Magnitude. Polaris 2.05.	Average Deviation in Magnitude.	Date, 1880 +	Mag. Argel. Uran.
		h. m.	o ,				
26	δ................	18 7.8	3 23	4.54	0.03	2.708	4...5
27	24	18 11.5	3 0	5.87	.01	5.210	6

VIRGO.

1	ω	11 32.8	81 15	5.69	0.06	4.332	6
2	ξ	11 39.6	81 8	5.21	.06	...	5...4
3	ν	11 40.2	82 51	4.31	.09	2.360	4...5
4	Λ¹	11 42.3	81 9	5.30	.02	4.368	6
5	β	11 44.9	87 36	3.58	.06	...	3 ..4
6	Λ²	11 49.4	80 57	5.72	0.04	4.368	6
7	b...............	11 54.3	85 44	5.36	.11	4.376	6
8	π	11 55.2	82 46	4.62	.08	5.327	4...5
9	o......	11 59.6	80 39	4.29	.08	2.363	4
10	10	12 4.0	87 29	6.33	.06	4.390	6
11	11	12 4.4	83 35	5.70	0.04	4.368	6
12	12	12 7.8	79 7	5.81	.04	4.365	6
13	L. 22954	12 9.5	99 38	5.95	.05	4.365	6
14	η...............	12 14.3	90 3	3.82	.03	5.327	3...4
15	16	12 14.8	86 4	5.32	.06	4.390	5
16	21	12 28.1	98 51	5.67	0.06	4.368	6
17	25	12 31.1	95 14	5.93	.06	4.368	6
18	L. 23608	12 32.5	86 6	6.12	.09	4.365	6
19	P. xii. 142........	12 32.8	87 32	6.06	.06	4.376	6
20	χ	12 33.6	97 23	4.67	.04	4.393	5

No. 1. Called red in the Uranometria Argentina. Spectrum III a !!! (Vogel.)
No. 2. 5.21 is the mean of two determinations (5.31, 5.11) made on 1882.360 with twenty extinctions, and on 1885.327 with six. **H.P. = 4.92.** Spectrum I a ! (Vogel.)
No. 3. Spectrum III a! (Vogel.)
No. 4. Spectrum I a ! (Vogel.)
No. 5. 3.58 is the mean of eight determinations made on as many nights, three at Oxford and five at Cairo, involving ninety extinctions made between 1882.363 and 1883.169. H.P. = 3.72.
Nos. 7 and 18. Spectrum I a ! (Vogel.)
No. 8. The determination made on 1882.363 (see Memoirs R.A.S., vol. xlvii, p. 455) is rejected. The observations were interrupted by clouds. Spectrum I a ! (Vogel.)
No. 10. The estimates of magnitude by various authorities differ. (See Uran. Argent., p. 318.)
No. 12. Olbers suggested variability from comparisons with Vesta. (See Berl. Jahr., p. 197.)
No. 14. The determination made on 1882.363 (see Memoirs R.A.S., vol. xlvii, p. 455) is rejected. See note to No. 8. Spectrum I a ! (Vogel.)
No. 19. Spectrum III a !! (Vogel.)

Reference Number.	Star's Designation.	R.A. 1890.	N.P.D. 1890.	Adopted Zenithal Magnitude. Polaris 2.05.	Average Deviation in Magnitude.	Date, 1880 +	Mag. Argel. Uran.
		h.　m.	°　′				
21	γ	12 36.1	90 51	2.07	0.12	2.363	3.. 2
22	ρ..................	12 36.3	79 9	4.95	.09	4.401	5
23	31	12 36.4	82 35	5.80	.05	4.376	6
24	32	12 40.1	81 44	5.38	.04	4.365	6
25	33	12 40.8	79 51	6.08	.07	4.376	6
26	41	12 48.3	76 59	6.10	0.02	4.376	6
27	ψ	12 48 6	98 56	5.12	.09	4.401	5
28	δ...................	12 50.1	86 0	3.47	.10	2.363	3
29	44	12 54.0	93 13	5.73	.06	4.368	6
30	46	12 54.9	92 47	5.97	.05	4.365	6
31	ε...................	12 56.7	78 27	3.01	0.07	2.363	3...2
32	48	12 58.2	93 4	6.42	.04	4.376	6
33	B.A.C. 4394	13 2.8	98 24	5.86	.09	4.390	6
34	θ...................	13 4.2	94 57	4.49	.04	5.327	4...5
35	L. 24586	13 8.9	78 6	5.73	.05	4.390	6
36	59	13 11.3	80 0	5.17	0.07	4 393	5
37	P. xiii. 41	13 11.8	75 45	5.76	.02	4.401	6
38	σ	13 12.1	83 57	5.14	.04	4.365	5
39	L. 24703	13 13.2	85 46	6.19	.03	4.401	6
40	L. 24708	13 13.5	85 44	0.53	.02	4.390	6
41	B.A.C. 4470	13 16.1	87 20	5.96	0.02	4.376	6
42	64	13 16.6	84 16	5.97	.05	4.390	6
43	65	13 17.6	94 21	6.13	.04	...	6
44	66	13 18.8	94 35	5.95	.08	4.376	6
45	α...................	13 19.4	100 36	+ 0.04	.03	...	1

Nos. 21, 22, 23, and 24.　Each has the Spectrum I a ! (Vogel.)

No. 21.　Σ 1670.　Binary.　Period 185 years.　The relative brightness of the two components is variable.　(See Mens. Mic., Introduction, p. lxxii, and Pulkova Observations, vol. ix, p. 122.)

No. 27.　Called red in the Uranometria Argentina.

No. 28.　Birm., No. 297.　The colour was noted to be yellow.　Spectrum III a ! ! ! (Vogel.)

No. 29.　Σ 1704, Dist. 21″.　The brighter star observed.

No. 34.　Σ 1724.　Triple.　The close pair, dist. 7″, observed as one mass.　The determination made on 1882.363 (see Memoirs R.A.S., vol. xlvii, p. 455) is rejected.　See note to No. 8.

No. 38.　Birm., No. 302.　The star is decidedly of an orange tint.　Spectrum III a ! ! (Vogel.)

No. 42.　5.97 is the mean of two determinations (5.94, 6.00) made on 1884.390 with twenty extinctions, and on 1885.352 with six.　H.P. = 5.68.

No. 45.　+ 0.04 is determined from three nights' observations at Cairo alone, made between 1883.127 and 1883.136.　Further, observations on four nights at Oxford, between 1882.363 and 1883.207, when corrected for mean atmospheric absorption, gave + 0.19.　H.P. = 1.23.　For explanation of notation see Preface.

Reference Number.	Star's Designation.	R.A. 1890.	N.P.D. 1890.	Adopted Zenithal Magnitude, Polaris 2.05.	Average Deviation in Magnitude.	Date, 1880 +	Mag. Argel. Uran.
		h. m.	° ′				
46	70	13 23.1	75 38	4.97	0.04	4.401	5
47	71	13 23.8	78 37	5.93	.04	...	6
48	74 ....:.. ..	13 26.3	95 41	4.97	.06	4.363	5
49	76	13 27.2	99 36	5.62	.07	4.368	5
50	78	13 28.5	85 47	5.11	.05	4.376	5
51	ζ.............	13 29.2	90 2	3.36	0.06	2.363	3 ...4
52	80	13 29.8	94 50	5.91	.07	4.390	6
53	82	13 35.8	98 9	5.54	.05	4.390	6
54	84	13 37.5	85 54	6.00	.05	...	6
55	P. xiii. 174........	13 38.2	94 57	6.05	.05	4.393	6
56	B.A.C. 4591	13 41.4	99 9	6.21	0.03	4.393	6
57	88	13 42.5	96 13	6.07	.04	4.393	6
58	90	13 49.0	90 59	5.39	.05	4.390	6...5
59	P. xiii. 238........	13 49.2	97 31	6.58	.03	4.393	6
60	ν................. ..	13 49.9	99 1	6.73	.08	4.398	6
61	92	13 50.8	88 25	5.97	0.03	4.393	6
62	τ.........	13 56.1	87 55	4.29	.07	2.363	4
63	95	14 0.9	98 47	5.85	.02	4.390	6
64	P. xiv. 12	14 6.7	87 4	5.44	.05	2.363	5...4
65	κ.................	14 7.0	99 46	4.09	.10	2.368	4...5
66	ι.................	14 10.2	95 28	3.94	0.04	2.368	4
67	W.B. xiv. 143 ...	14 10.8	92 41	6.02	.05	4.401	6
68	λ	14 13.3	102 52	...	.03	2.368	5...4
69	υ.................	14 13.8	91 45	5.00	.04	4.401	5
70	L. 26200	14 14.1	89 6	6.31	.02	4.393	6

No. 47. 5.93 is the mean of two determinations (5.99, 5.87) made on 1884.393 with twenty extinctions, and on 1885.352 with six. H.P.= 5.66.

No. 48. Birm., No. 309. Called red in the Uranometria Argentina. The colour is orange.

No. 50. Spectrum I a! (Vogel.)

No. 51. Spectrum I a !! (Vogel.)

No. 54. Σ 1777, Dist. 3″. Observed as one mass. 6.00 is the mean of two determinations (5.98, 6.02) made on 1884.365 with twenty extinctions, and on 1885.352 with six. H.P.=5.73.

Nos. 57 and 60. The notation is that of Argelander; No. 57 is called ν in B.A.C.

No. 63. Called red in the Uranometria Argentina.

No. 64. Probably variable within small limits. See Uran. Argent., p. 320. H.P.=4.97.

No. 65. Birm., No. 322. The red colour is not salient.

No. 68. The magnitude, when corrected for mean atmospheric absorption, is 4.70.

Nos. 69 and 70. Gould suggests variability. (See Uran. Argent., p. 320.)

Refer-ence Number.	Star's Designation.	R.A. 1890.	N.P.D. 1890.	Adopted Zenithal Magnitude. Polaris 2.05.	Average Deviation in Magnitude.	Date, 1880 +	Mag. Argel. Uran.
		h. m.	° ′				
71	L. 26289	14 17.6	88 15	6.40	0.02	4.393	6
72	φ	14 22.6	91 44	4.87	.04	4.365	5
73	106.................	14 22.8	96 24	6.02	.06	4.401	6
74	B.A.C. 4798	14 24.2	88 41	6.00	.04	4.401	6
75	μ	14 37.3	95 11	3.80	.05	2.368	4
76	108.................	14 39.9	88 49	5.71	0.05	4.393	6
77	109.................	14 40.7	87 38	4.04	.06	...	4...3
78	L. 26980	14 43.4	83 34	6.30	.04	4.401	6
79	W.B. xiv. 880 ...	14 48.2	83 18	6.87	.02	4.393	6
80	L. 27162	14 50.0	82 45	6.93	.08	4.390	6
81	L. 27297	14 53.9	84 59	6.31	0.02	4.393	6
82	110.............. ...	14 57.3	87 27	4.99	.06	...	5
83	L. 27541	15 2.2	84 3	5.90	.02	4.393	6

No. 72. Σ 1846, Dist. 5″. Observed as one mass.
No. 73. Called red in the Uranometria Argentina.
No. 74. Spectrum I a ! (Vogel.)
No. 77. 4.04 is the mean of two determinations (4.14, 3.94) made on 1882.368 with twenty extinctions, and on 1885.352 with six. H.P. = 3.72.
No. 78. Gould thinks it impossible that the star could have been so faint when seen by Argelander. (See Uran. Argent., p. 321.) Spectrum I a ! (Vogel.)
No. 82. 4.99 is the mean of two determinations (4.92, 5.06) made on 1884.393 with twenty extinctions, and on 1885.352 with six. H.P. = 4.56.

VULPECULA.

1	Bradley 2409......	19 1.5	65 55	5.43	0.02	3.719	6
2	W.B. xix. 17	19 3.3	68 29	6.06	.04	3.877	6
3	B.A.C. 6574	19 7.9	68 38	5.68	.03	3.719	6
4	1.................	19 11.5	68 48	4.83	.10	2.601	5...4
5	2.................	19 13.1	67 10	5.26	.06		6
6	3.................	19 18.4	63 57	5.05	0.05	3.719	5...6
7	4.................	19 20.6	70 25	5.04	.06	3.730	5
8	5.................	19 21.4	70 7	5.63	.03	3.719	6
9	6.................	19 24.1	65 33	4.58	.09	...	4...5
10	9.................	19 29.7	70 28	5.12	.08	3.711	5

No. 5. Double. Dist. 1.″8. Observed as one mass. 5.26 is the mean of two determinations (5.29, 5.24) made on 1883.692 with twenty extinctions, and on 1885.442 with six.
No. 9. Σ 42¹, Dist. 396″. The distant companion not observed. 4.58 is the mean of two determinations (4.57 and 4.59) made on 1882.601 and on 1882.604, each with twenty extinctions.

Reference Number.	Star's Designation.	R.A. 1890.	N.P.D. 1890.	Adopted Zenithal Magnitude. Polaris 2.05.	Average Deviation in Magnitude.	Date, 1880 +	Mag. Argel. Uran.
		h. m.	° ′				
11	10	19 39.1	64 30	5.20	0.13	3.721	6
12	12	19 46.3	67 40	5.49	.09		5
13	13	19 48.8	66 12	4.97	.11	...	5...4
14	Bradley 2541	19 49.9	65 58	5.36	.04	3.719	6
15	14	19 54.5	67 12	5.94	.05	3.721	6 ..5
16	15	19 56.6	62 33	5.03	0.04	3.721	5
17	Bradley 2559	19 57.1	65 30	6.09	.06	3.840	...
18	16	19 57.4	65 22	5.37	.03	3.840	5
19	17	20 2.2	66 42	5.26	.08	3.719	5 ..6
20	18	20 6.0	63 25	5.45	.03	3.721	6
21	19	20 7.2	63 31	5.54	0.11	3.730	6
22	20	20 7.4	63 51	5.72	.08	3.721	6
23	21	20 9.7	61 38	5.30	.06	3.719	6...5
24	B.A.C. 6966	20 10.6	64 45	5.09	.10	...	5
25	22	20 10.8	66 50	5.45	.09	3.719	6
26	23	20 11.2	62 31	4.72	0.07	3.730	5
27	24	20 12.1	65 40	5.67	.14	3.719	6
28	25	20 17.3	65 54	5.34	.12	3.730	6
29	L. 39329	20 20.8	68 57	6.01	.04	3.738	6
30	27	20 32.4	63 55	5.73	.09	3.721	6...5
31	29	20 33.6	69 11	4.68	0.03	3.738	5
32	28	20 33.7	66 16	5.33	.04	3.719	5...6
33	30	20 40.1	65 7	5.02	.02	3.721	6...5
34	31	20 47.4	63 19	4.62	.02	3.730	5
35	32	20 49.9	62 22	5.31	.05	3.738	5...6
36	33	20 53.4	68 6	5.49	0.04	3.738	6...5
37	P. xxi. 120	21 19.7	64 18	5.84	.08	3.719	6
38	35	21 22.8	62 52	5.34	.04	3.738	6...5

No. 12.   5.49 is the mean of two determinations (5.45, 5.53) made on 1883.719 with twenty extinctions, and on 1885.442 with six. H.P. = 5.02.

No. 13.   4.97 is the mean of two determinations (5.09, 4.88) made on 1883.730 with twenty extinctions, and on 1885.442 with six. H.P. = 4.66.

No. 21.   Birm., No. 537. The red colour is not salient.

No. 24.   5.09 is the mean of two determinations (5.06, 5.12) made on 1883.730 with twenty extinctions, and on 1885.442 with six. H.P. = 4.76.

No. 26.   Birm., No. 546. The colour of the star is yellow.

No. 35.   Variable? (See Gilliss, Ast. Obs., p. 670.)

www.ingramcontent.com/pod-product-compliance
Lightning Source LLC
Chambersburg PA
CBHW030602270326
41927CB00007B/1020